The Art and Craft
of Natural Dyeing

The Art and Craft of Natural Dyeing

Traditional Recipes
for Modern Use

J. N. Liles

The University
of Tennessee Press

Knoxville

Library of Congress Cataloging in Publication Data

Liles, J. N., 1930–2002
    The art and craft of natural dyeing : traditional recipes for
modern use / J. N. Liles—1st ed.
        p.   cm.
    Includes bibliographical references (p.)
    ISBN 0-87049-669-7 (cl.: alk. paper)
    ISBN 0-87049-670-0 (pbk.: alk. paper)
    1. Dyes and dyeing, Domestic. 2. Dyes and dyeing—Textile fibers.
    I. Title. II. Title: Natural dyeing.
TT854.3.L55   1990
667'.26—dc20   90-12045   CIP

To Dale —
wife, advisor, critic, and best friend

We know better where we are going if we
understand where we have been.

Contents

7   Purple Dyes   155

8   Orange Dyes   167

Appendixes

Glossary

References

Index

Illustrations

Preface

My interest in natural dyes originated in the early 1960s when my wife, Dale, started working with them. A month after purchasing a copy of Emma Conley's *Vegetable Dyeing,* we attended the October Fair of the Southern Highland Handicraft Guild in Gatlinburg, Tennessee, where Mary Frances Davidson, author of *The Dyepot,* was demonstrating dyeing techniques. The effect on Dale was immediate: in fact, she was collecting broom-sedge, blackberry briars, and mountain dog hobble the following week-end. Dale has worked continuously with natural dyes since that time, making particularly good use of their unique properties when she started spinning and when, a few years later, she began working with handmade felt articles.

Until 1977 my use of natural dyes was confined to the biology classroom and laboratory and had not extended to the dyeing of fibers. Carmine (cochineal), alizarin (madder), and haematoxylon (logwood), all fiber dyes used by the ancients, had been borrowed by the biologist many years ago for use as animal and plant tissue stains. The same is true of certain of the early synthetic fiber dyes such as neutral and Congo reds, crystal violet, Bismark brown, safranine, and methylene blue. In fact, certain of the early synthetic fiber dyes are currently available *only* because they are still used by biologists.

In 1977 I joined a historical reenactment group interested in studying the period between 1769 and 1800 in the area that is now East Tennessee. My research soon revealed that the early settlers in this area used a great deal of cellulosic fiber (linen, fustian, and cotton) for clothing, some of which was dyed. I wished to experiment with dyeing linen clothing with natural dyes, but found that twentieth-century natural dye manuals emphasize wool and contain few recipes for the cellulosic fibers. Generally speaking, one can find such detailed recipes and technologies only in pre–twentieth-century works. Thus, I decided to fill this void and to concentrate, especially, on collecting traditional recipes for cotton and linen dyes. Encouragement in this direction was immediately forthcoming from Mary Frances

Davidson and Fred Gerber, both well known in the field of natural dye technology. At this point, then, I started what was to become the most interesting and rewarding research of my career: collecting, experimenting with, and passing on the lore of traditional dyeing. I also feel I am performing a worthwhile service by transmitting what we know of these ancient techniques. After all, natural dyes were used for several thousand years, synthetic dyes only since 1856. The former are a valuable part of our heritage!

My goals in this book are several:

1. To reestablish and maintain continuity with the past, so that as our ancient traditional textiles age, fade, and deteriorate, new ones comparable to the originals may be produced.

2. To add to the craftsperson's color-shade palette. Many of the natural dyes, premixed by nature, present a different and often more pleasing visual effect than do modern synthetic dyes. In former times often two or three natural dyes or a natural dye and an early synthetic, as well as two or more mordants, were combined, just to yield that special visual effect. Similar results may be obtained with our modern synthetics, but because of the cost this is all too seldom undertaken.

3. To celebrate the colors of the ancient world and to illustrate that they were not always muted, but could also be bright and vibrant, yet with a distinct softness. Just observing a well-preserved or much-used old quilt containing Turkey red, chrome green, black oak bark yellow, and indigo will prove the point, not to mention seeing the brilliant hues of scarlet red coats, Persian rugs, Indian palempores, and Gobelin tapestries, some existing specimens of which date to the thirteenth and fourteenth centuries.

4. To extend research in use of the traditional dyes.

5. To compile a relatively complete description of the mineral dyes much used in the late eighteenth and nineteenth centuries, particularly on cotton and linen, some on wool and silk.

6. To provide recipes for use in resist-dyed craft fabrics. Many of the natural dyes penetrate less readily than do the modern synthetics, making them ideal for some of this work.

7. To provide and retain recipes from the original old literature, much of which is becoming more and more inaccessible.

8. To provide details of technique, so important to good work, particularly with piece goods.

Acknowledgments

First, I should like to give special thanks to my wife, Dale, for being the first to encourage me to try my hand at natural dyeing and for not objecting to having dye experiments going on in her studio literally every night during the first years of my study. I thank her also for the use of her library and for designing and sewing four traditionally dyed shirts for me, two in Turkey red.

Throughout the study, assistance and support have been continuous from Mary Frances Davidson and Fred Gerber. These remarkable people have most generously supplied me with their expertise, dyestuffs difficult to obtain, and numerous original books and manuscripts from the eighteenth and nineteenth centuries.

Special thanks to Penny Tschantz for essential editorial assistance, Cheryl Lynn for typing the manuscript, and The University of Tennessee Press for publishing the work.

I also wish to thank the members of the "Thursday Bee" of the Smoky Mountain Quilters who have constructed numerous quilts, wall hangings, and a vest from my traditionally dyed cotton piece goods. Two of the quilts are true heritage works – constructed in every detail as they would have been done in the late eighteenth and nineteenth centuries. Members of this group are Eva E. Kent, Linda Claussen, Sandra Cartwright, Rebecca Harriss, Ginger Neusel, Irene Wylie, Lessie Werner, Jean Lester, Carole Whitehouse, Renna Kent, Kristin Steiner, Willie Cronin, Merikay Waldvogel, Bets Ramsey, and Joyce Tennery.

Finally, the following have provided assistance at some point, and I wish to express my gratitude to all.

Rita Adrosko	Sally Garoutte
Hugh Bailey	Harriet Gill
Barbara Brackman	Mattiebelle Gittinger
Rita Buchanan	Harold Glass
Helen Cobb	Marian G. Heard
Janet DeBoer	David Hedberg

Laurel Horton
Mary K. Jarrell
Joseph Jones
Dr. Carl Kuttruff
Dr. Jenna Kuttruff
David Liles
Russell Liles

Dr. Kenneth Monty
Bette Raymond
Dr. George Schweitzer
Mary Stock
Sadye Tune Wilson
Roger Wolfe

The men of whom mention has been made in these pages are those who created the tints of the marvelous Cretan costumes, who harmonized the striking apparel of the Egyptians, enriched the sumptuous tunics of the Assyrians, produced the scarlet of the mantle of Christ, whitened the Roman togas, and made joyous the beautiful dresses of the Florentine ladies and those of the court of Burgundy.

These color magicians gave splendor to the Renaissance corteges in St. Mark's Square in Venice and to the XVI century parades of the capitals of Europe, made splendid the red shirts of the soldiers of Garibaldi and elegant the silhouettes of the ladies of the belle-epoque.

Franco Brunello
The Art of Dyeing

1 Introduction

Brief History of Dyeing

Color and color use comes over the horizon from prehistory into history in all cultural groups and on all major land masses. Although basic human needs fall into the three categories of food, shelter, and clothing, the practice of burying the dead on red ochre mounds as far back in time as the Old Stone Age (30,000 B.C.) and the use of mineral-colored earths in cave paintings as early as 15,000 B.C. strongly suggest that early humans were artistic as well as practical. Using color appears to have filled a complex need, spiritual and otherwise, from our very beginnings. Humans have used color for as long as we have sought shelter, certainly as long as we have produced textiles and clothing. Regardless of the culture, use of color was, and is, universal.

When did the dyeing of animal skins, basketry materials, and textiles actually start, and what were the first dyes? The answers to these questions are unknown, but the practice must have begun quite early. Certainly the fibers were available. Animal fur and wool was always at hand; wild cotton was indigenous to the American Southwest, Mexico, Central and South America, Africa, Persia, and India. Linen (flax) was somewhat more limited to Europe, Russia, the Mediterranean, and Egypt. The Indians of Mexico cultivated cotton for textile purposes as early as 5000 B.C. (Hochberg, 52). Fragments of cotton textiles dyed by complex processes, dating to approximately 2000 B.C., were unearthed in the excavations of Mohenjo Daro in the Indus Valley of India.

The first colors used for textiles were probably little more than stains with the exception of stable iron rust yellows and reddish oranges, bark tannin ochres and light browns, and iron tannate grays and blacks. These stable dyes remained in limited use into the beginning of the twentieth century. Bright yellows and yellow oranges from turmeric, saffron, and annotto, and pinks and rose pinks from safflower were undoubtedly used quite early also, because these dye directly, without any pre-treatment of the fiber other than washing.

Many ancient civilizations possessed excellent empirical dye technologies, as is evidenced by Egyptian and Peruvian tomb remains. By 1500 B.C. the Phoenicians had a thriving Tyrian (royal) purple dye industry in Tyre and other cities. Among the ancients, India was probably the most advanced. The Indians dyed all natural fibers well, especially the more complicated and time-consuming cotton (Gittinger, 16). Indian supremacy in cotton dyeing probably resulted from the country's early settlement, its large population and labor force, and the presence there of wild cotton, the best of the dye-bearing plants, and of the necessary natural metal mordants. In addition, the Indians' unusual love of color and patient, perfectionistic approach to life probably contributed to their preeminence among the ancients. In Europe, the Italian dyers, particularly those in Venice and Tuscany, were acknowledged as the best from the time of the Roman Empire through the sixteenth century (Brunello, xvii).

Until 1856, and much later in certain places and with specific dyes, all dyestuffs were of animal, vegetable, or mineral origin. However, in 1856 William Henry Perkin, a young English chemist, while attempting to synthesize quinine from aniline, a coal tar by-product, accidentally produced and discovered mauve, the first synthetic dye. Following Perkins' discovery, use of the traditional dyes of previous millenia declined increasingly until, by about 1915, little remained in use by industry or home craftspersons. Indeed, by the 1880s every country store in North America carried the new Diamond synthetic dyes for home use.

Traditional methods of natural dyeing were reintroduced by the craft revivals of the 1920s, mostly by spinners, weavers, and knitters, untrained technically in dye chemistry, for use in their finished products. The general unavailability of eighteenth- and early nineteenth-century dye manuals, as well as the scarcity of certain chemicals and exotic (imported) dyestuffs rather limited craft dyeing in this country and elsewhere to the less-complicated processes required for wool. In actuality, the majority of late eighteenth- and early nineteenth-century dye manuals contain more recipes for cotton than for wool or silk, but even in earlier times there was far less home dyeing of cotton and silk than of wool. This situation has largely held to the present; little cotton, linen, or silk has been dyed by traditional methods in North America for nearly a hundred years. The one notable exception is indigo dyeing, which has remained in continuous use on all of the natural fibers.

How Good Were the Pre-Synthetic Era Dyes?

Many of the natural dyes were excellent, particularly when produced in an established dyehouse; plenty of museum specimens, old quilts, coverlets, etc., exist as proof. Specialization in natural dyeing started very early. In ancient India indigo blue dyeing was virtually always done by an expert, and Turkey red dyed cotton was usually produced in dyehouses set up for that purpose alone. Home dyeing varied from excellent to poor, depending upon the dedication, knowledge, and experience of the dyer.

The main reasons for shifting from natural to synthetic dyes, initially, were economics and brilliance of color. It is much less expensive to dye to desired color, intensity, and shade with a pure synthetic dye than to process natural materials with varying dye content. Many of the early synthetic dyes were actually inferior to the established natural dyes in light- and washfastness but were adopted anyway because of economics and extension of color range. The best of our modern synthetic dyes are superior in light- and washfastness to the natural dyes, with a few exceptions, but many items we purchase today are not dyed with the best modern dyes, and are not particularly fast, partly because certain hues and shades are not possible with the most durable dyes.

The natural dyes are not stark, pure colors like the synthetics. Instead, they have been pre-mixed by nature, thus rarely clashing when placed side by side. This imparts to them some of their special beauty. Mixing modern dyes can produce some of this effect, but is all too seldom done. And the natural dyes possess additional beauty because they come from living things. (I sometimes feel that some of that life is still there.)

After having spent the past decade intensively researching and reproducing the traditional pre-synthetic era dyes, I have reached the conclusion that they were, and are, very good. In fact, many of them far surpass our modern dyes in sheer, subtle beauty, if not always in fastness.

Traditional Dyeing: General Features

Dyeing differs from painting and staining, the two processes closest to it. Dyeing involves the chemical combination of a part of a dye molecule with a chemical grouping in the fiber molecule, or bridging by means of a "mordant" molecule which combines with both the fiber and dye molecules. Mordants include metal oxides, tannins, and oxyfatty acids. Staining does not involve chemical combination, and usually quickly washes out. Painting involves the application of insoluble pigment, suspended in a liquid binder, to a surface.

Natural fibers suitable for yarn, piece goods, basketry materials, etc., fall into two categories: the *protein fibers* (wool, silk, skin, horn, and feathers), and the *cellulosics* (cotton, linen, hemp, ramie, jute, and fibrous vegetable basketry materials). Proteins are long chain polymers of amino acid molecules; cellulose is composed of long chain polymers of glucose (simple sugar) molecules.

Traditional dyes also fall into two categories. *Mineral dyes* are *inorganic*, like many painting pigments, though the material is precipitated more within the fiber than is the case with painting pigments. The first mineral dye (and possibly our oldest dye) was hydrated iron oxide (rust). On the other hand, *traditional dyes* derived from animals and plants are *organic*. The vast majority of the organic dyes, by themselves, will only stain fibers. To effect chemical combination (dyeing) a mordant or mordants are necessary. The mordant combines chemically both with a grouping of the fiber molecule and with the dye. Thus the mordant acts as a chemical bridge between fiber molecule and dye molecule. The chemistry of mordant action is somewhat complex, often involving coordinate covalent bonding, and the interested reader with knowledge of organic chemistry is referred to the discussion by Trotman (430–43) and Baker (207–27).

Dyes that require mordants are called *adjective dyes* and have been in use at least as far back as 2000 B.C. (Indus Valley of India and elsewhere). Dyes which combine without mordants, or which contain their own natural mordant (usually tannins), are called *substantive dyes*. Examples of substantive dyes include turmeric (yellow), saffron (yellow-orange), safflower (rose and pink), annotto (orange), black walnut (brown), and cochineal (on wool only, crimson). Some, such as cochineal and black walnut, are sometimes used with and sometimes without a mordant, depending upon the desired color. Indigo and Tyrian purple also do not require or use mordants, but these are "vat dyes" which operate by a different principle.

Early on it was thought that the mordant opened the pores of the fibers so that the dye could enter. This theory, however, did not answer the observation that a number of the traditional dyes are *polychromic;* i.e., several colors or shades result from using the same dye with different mordants. Cochineal and madder are excellent examples.

Macquer (a French scientist) proposed in the 1750s the correct theory that a chemical combination took place between mordant and dyestuff and fiber (Brunello, 231). With adjective dyes, the color is primarily the result of the metal-dye complex.

Inclusions and Omissions

This book contains a brief history, with references, a chemistry, and a botanical derivation (where applicable), as well as detailed recipes for each dye. In general, each recipe is referenced, unless I have greatly modified it. Some of the recipes are my own. In the past, many dye manuals have omitted references for recipes, and I consider this improper. Indeed, some nineteenth-century and early twentieth-century authors were the worst thieves of all, often lifting recipes word for word without even testing them. I have tested all but two or three of the recipes at least once. All recipes have been tested unless specified.

The guilds of the Middle Ages and earlier groups kept dye knowledge pretty well secret, so that little information was written down. Finally, in the early 1500s a German dye manual appeared (Matthews, 15), and this was followed by the very important *Plichto* by Rosetti (Venetian) in 1548, which is now available in the original Italian and in translated form. From this point on, the ice seemed to have been broken, and between 1700 and 1885, probably fifteen or so excellent books were published; of these, the two volumes by Edward Bancroft are particularly significant. This outpouring, of course, accompanied and was a part of the tremendous development of the science of chemistry, particularly from 1750 on. Through my wife Dale's library, the library and interlibrary loan service of the University of Tennessee, the library of the Southern Highland Handicraft Guild, and through materials lent or given to me by Mary Frances Davidson and Fred Gerber, I have managed to obtain and consult most of this very important and rather inaccessible material. One very significant recent source of reference materials is Brunello's *The Art of Dyeing in the History of Mankind,* made available in English translation by the Phoenix Dye Works, Cleveland, Ohio.

Many of the recipes in this book are modified from the older sources (1750–1885) because it has been my experience that the best recipes come from the period of peak use of each particular dye, and in some cases, methods were cheapened as demand for a product increased.

A number of the recipes use the ancient, time-tested, exotic dyes. Many of these, properly used, have the best reputations for light- and washfastness. Also, most eighteenth- and nineteenth-century American dyers were trained in Europe and continued using tested, time-honored materials rather than experimenting with local plants. Like me, they wanted to be sure that the fabrics they provided would not fade noticeably in short order. Much dyeing is ruined by poor technique. Therefore, I have made an especial effort in this direction, particularly with piece goods, which require more care than yarn.

The book contains no recipes with lichen and mushroom dyes. This omission is simply due to the fact that I have not experimented with them, and they do not have the reputation as having been used on the cellulosic fibers. I have also omitted recipes for many beautiful shades and colors because they possess such poor lightfast characteristics. Tin-mordanted logwood purples on cotton are prime examples.

This book was originally conceived to be restricted to cotton and linen but as the research progressed my fascination with silk had to be satisfied, and I felt that, for comparative purposes, wool should be included as well. Therefore, recipes for all fibers are present.

2 Dyeing Procedures

General Dyeing Techniques

The difference between mediocre and very good dyeing often hinges on paying attention to what may appear to be minor details. Attention to detail is of paramount importance, particularly to beginning dyers, until certain procedures become matter of course. It is hoped that the following suggestions will prove helpful:

1. For most uniform results for a project (e.g., 3 pounds of dyed yarn for a sweater), (a) use yarn from the same lot, (b) scour all of the yarn at the same time, and in the same vessel, (c) mordant all in the same vessel at the same time. Follow the same procedure in the dyeing. Thus, the entire lot will be treated identically.

2. Many of the natural dyes are pH sensitive (change color, tint, or shade in acidic or alkaline solution). For this reason a number of the recipes call for the addition of small amounts of acids or alkalis. Sometimes acids or alkalis are necessary only if the water is extremely hard or soft.

3. Careful maintenance of proper mordanting and dyeing temperature and duration significantly affects resulting color and quality of many dyes. It is suggested that you read the entire recipe, including any notes or additional comments, before you start the dyeing process.

4. If bright, clear colors are desired, always use scrupulously clean nonreactive or copper or brass vessels. Iron vessels will dull (sadden) bright, clear colors. Nonreactive vessels include stainless, enameled, glass, glazed earthenware, and plastic containers. Chipped enameled vessels should be repaired or iron staining may result. Of course, earthenware and plastic cannot be used for high temperatures. Copper and brass are reactive, but not enough copper is released to affect the color appreciably.

5. Level dyeing is much more difficult with piece goods than yarn. Thus, more careful mordanting and dyeing is required with piece

goods. It is especially important to rinse or wash newly dyed material (indigo excepted) thoroughly before hanging to dry; a nonabsorbent plastic line is recommended. Clothespinning piece goods, unless they are very heavy, is most highly recommended.

6. The vast majority of the older dye recipes recommend drying the dyed articles in the shade, out of direct sunlight. Exceptions to this general principle sometimes include articles dyed in madder and natural indigo. Perhaps direct sunlight more quickly destroys the natural materials which dull the effect of the alizarin in the madder or the indigotin in the natural indigo. It is true that madder (Turkey red) dyed yarns were exposed to strong sunlight in the past. On the other hand, direct sunlight has a deleterious effect on scoured wool. My experience compels me to suggest drying virtually all dyed articles out of direct sunlight. It is also probably best to follow the same recommendation with mordanted items to be dried, particularly if chrome mordanted.

7. Never start a large project or one involving expensive materials without running test samples first. Wools, silks, cottons, and linens differ. Some dye well and level while others may not. Some require more or less mordant and dyestuff to achieve good or desired color. Use inexpensive goods for experimentation and development of good technique. Rayon, old cotton sheeting, cotton cloth, and inexpensive cotton yarn are ideal for cotton and linen dyes. Some commercial woolen yarns and fleece are also inexpensive. Once good technique and confidence have been attained, you may use expensive yarns, fleece, and piece goods. Generally speaking, expensive goods yield the best colors.

8. Start off with yellows, off-yellows, browns, greys, logwood blues, and cochineal reds, crimson, and purples, using relatively simple recipes. Later, as you develop skill, you should add indigo blues, overdyes, and madder reds. Specialization may be in order. For example, some may wish to deal only with the many aspects of indigo or other blue dyeing.

9. Mixing several dyes and mordants can result in just that highly desired subtle shade, so necessary for a particular project. Experience and effort can pay off in a big way.

Materials and Equipment

Utensils

In former times the best vessels for hot scouring, mordanting, and dyeing were iron for dull or sad colors, and copper, brass, and block tin for bright colors. Glass and crockery were often used for cold or warm mordanting and dyeing, and wooden barrels were sometimes used for indigo. As copper and iron have always been quite expensive, they were replaced a number of years ago by enameled vessels. Enameled vessels are nonreactive, thought difficult to clean, particularly if used with logwood or chrome mordant, and the enamel layer is so thin in vessels of new manufacture that the slightest jar chips it, permitting rust to start. Therefore, if possible, start off with good-quality stainless steel vessels. The initial cost will be greater, but in the long run stainless will prove economical since it lasts indefinitely, can be used with all dyes and mordants, and can be cleaned of anything with steel wool pads. I further suggest that the stainless vessel have a capacity of at least 4 1/2 gallons if you expect to scour, mordant, or dye up to a pound of yarn at a time or 1/2 pound of piece goods. Plastic wastebaskets are inexpensive and excellent for scouring, mordanting, and dyeing where water temperatures are not expected to exceed 140° F. Widemouth glass jugs are also fine where only a small vessel and low temperature is needed, and earthenware crocks, though heavy, hold heat well, and may be used with hot (140° F or less) mordants and dyebaths. A few dyers use aluminum vessels, which result in colors somewhat different from those produced in nonreactive vessels. This effect is less with alum-mordanted items.

Water

For practically all natural dyes except madder, logwood, weld, and brazilwood, soft water is best. In former times, rainwater was considered ideal, riverwater next best, and well water the last choice since it often contained the largest amounts of dissolved salts. Not only do salts alter the color of some dyes, but they can sometimes cause spotting, particularly on piece goods. Iron contamination can really create havoc with bright colors. It is well to seek out a source of good water or to use deionized water if it is readily available. Rainwater, at present, probably contains contaminants. If iron contamination is suspected, dissolve a few crystals of potassium ferrocyanide in about one-half ounce of water. Add a little vinegar or weak acid. If the solution remains clear, it is iron free, but if the solution turns blue, iron contamination is present. The blue color results from formation of Prussian blue, which is iron dependent.

Work Area

Scouring, mordanting, and dyeing may be done outside, over an open fire if necessary, if a source of water is near. Most often, though, the work will be done inside. Ideally, the workroom should be well lighted and ventilated and should possess a heat source such as a hot plate or gas burner, a water source, and at least one rinsing or soaking sink. The room should also contain a work table, chair, and storage space or cabinets.

Additional supplies and equipment for the workroom include

1. detergent, paper towels, wastebasket;
2. stirring devices: glass rods, wooden dowels, wooden spoons;
3. cotton sheeting for labels;
4. waterproof laundry marking pen;
5. cotton string, for skein ties;
6. kitchen-type measuring spoons;
7. glass or nonreactive plastic measuring cylinders or jars;
8. rubber gloves;
9. strainer or cheesecloth;
10. thermometer;
11. gram or ounce scale;
12. pH paper;
13. plastic clothesline or drying rack;
14. glass or plastic jugs, for storing dyeliquors;
15. mortar and pestle;
16. plastic pails, for rinsing and cold mordanting and dyeing;
17. face mask.

Selection of Fiber

Advantages of cotton and linen as compared to wool:

• It is difficult to achieve with wool the beauty of cotton or linen dyed indigo, cutch chocolates, Turkey red, Napolean's blue, and logwood black, particularly if mercerized cotton (perle), high-quality linen, or high-quality cotton sateen or velveteen piece goods are used.

• Clothes moths are not a menace.

- Temperature shock is of no consequence (no felting). You may place cold cotton or linen into boiling water, or vice versa, if convenient or necessary.
- Cellulosic materials are stronger wet than dry and may be subjected to rough treatment.
- Cellulose is not damaged by strongly alkaline solutions, whereas wool and silk are quite adversely affected.
- Cellulose takes mineral dyes well with excellent lightfastness, general washfastness, and with less harshness than is the case with wool.
- Most dyebaths work best at relatively low temperatures (70° to 160° F). Silk, though a protein fiber, usually mordants and dyes more like cotton than wool. Proper dyebath temperature is usually between those required for cotton and for wool.

Disadvantages of cotton and linen as compared to wool:

- Scouring takes longer, and is more expensive.
- Mordanting is usually more involved, time consuming, and expensive.
- Cotton takes longer to dry and is heavy when wet.
- Tangled skeins are far more common with cotton than with wool (use 4 or 5 ties with skeins of cotton, and never use woolen ties).
- While the cellulosics are not affected by alkalis, they are by acids, particularly if they are allowed to dry on the material (the one exception is acetic acid or vinegar).
- Often, but not always, stronger dyebaths are necessary for cotton and linen to achieve an equivalent color level.
- In some cases, washfastness is not quite as good on the cellulosics. (Lightfastness, rubfastness, and sweatfastness are probably comparable.)

Fiber Preparation: Skeining

Skeining from a cone of yarn prior to scouring and subsequent dyeing is accomplished with a swift, weasel, niddy-noddy, or between the web of the thumb and first finger and elbow. One turn, all the way around, between the thumb and elbow will yield a length of about 3 feet. To tie the skein proceed as follows (Use cotton for the ties in all cases. Woolen skeins should have a minimum of 3 ties; cotton, linen, and silk should have a minimum of 4 ties.):

1. First tie the ends of the skeined-off yarn together. With one piece of cotton string, tie a *loose* loop around the skein where the ends are tied together, tying the cotton tie's knot into the knot of the skein.
2. Second loop: About one-third to one-fourth of the way around the skein, tie a second loose loop (not too loose, just loose enough for the wet yarn to expand without being crowded). Too loose a tie permits the ties to move too much; too tight a tie will result in a resist, particularly with natural dyes.
3. Third and fourth loops (a double loop, dividing the skein thickness in half): In the middle of the remaining untied areas of the skein, divide the yarn into two bunches. Tie a loop (loose) around one bunch, tie a non-slipping knot, encircle the remaining yarn with a loose loop, and tie a non-slipping knot. If two of these double ties are used, it is important to divide the threads in the skein so that those in the first loop in the first tie are in the first loop of the second tie, etc. (no cross-overs). Many books indicate use of figure 8 knots for the double ties. I do not favor this procedure because one side can pull tight and produce a tie-dyed effect. This is particularly true of cotton and linen yarns.

Safety Precautions

Most mordants and mordant assistants are poisonous, but, with care and common sense, they may be used safely. Although most of these materials are hazardous, they are no more so than common household materials such as lye, ammonia, gasoline, materials containing petroleum distillates, solvents containing acetone or xylene, insecticides, weed killers, nail polish remover, battery acid, certain paints, concentrated cleaning materials, medicines, bleach, disinfectants, etc. The fact is that we live in a world of manufactured chemical compounds, more so now than ever before. I recently visited a fabric store that smelled so strongly of formaldehyde and other permanent press chemicals that I wondered why anyone would work there. On the other hand, a certain paranoia exists at present which has reached somewhat absurd proportions. I read one article within the past three years which listed vinegar (5% acetic acid) and cream of tartar (present in grapes and collected from wine casks) as hazards comparable to toxic ammonia, chrome, tin, and copper mordants.

Rules of safety when using dyes and mordants:
- Do not use cooking vessels for mordanting or dyeing.
- Do not mordant in the kitchen.
- Dyeing is permissible in the kitchen *only* when you are using known nontoxic substantive dyes such as turmeric or saffron or other

nontoxic natural dyes, used substantively or with vinegar, as in Easter egg dyeing. Other dyes suitable for kitchen use would be onion skins, cochineal, red beet juice, tea, spinach, and red cabbage.

• Wearing a face mask is recommended when you are weighing and mixing fine chemical powders and dyes. This is especially important if the work is being done inside. Even if the dye or chemical is considered non-toxic, as is the case with many natural dyes, inhalation of dusts is best avoided.

• Mordanting and most dyeing should be done in a well-ventilated room or outside, particularly if the mordant or dyebath temperature exceeds 180° F. Solutions of metallic salts that do not decompose upon heating, such as chrome, tin, copper, alum, and iron, do not enter the air unless the solution boils. With boiling, the dissolved salt does enter the air in small amounts in fine water droplets, a process called "entrapment." However, poisonous vapors do not emanate from mordant baths kept well below the boiling point.

• Rubber gloves should always be worn when you are handling materials either in the mordant or dyebath. Even if the dyebath is not poisonous, the hands, and especially the fingernails, may remain stained or dyed for some time. It is especially important never to place bare hands in strong oxidizing agents such as potassium dichromate (chrome), potassium permanganate (purple crystals), compounds of lead, or strong acids or alkalis.

• Do not prepare tin or iron mordants where fine metal is dissolved (reacted) in strong acids unless you have had considerable training in chemistry. In any case, these "spirits" should be prepared outside or in a chemical hood only. The reason for including recipes for these mordants is that they were widely used in the past and because they often yield excellent results. It is not dangerous to dissolve iron filings in vinegar, the earliest method of producing liquid iron mordant.

• Never mordant while eating, drinking, or smoking, or when using poisonous dyes or dyes containing poisons. *Know and understand what you are working with.*

Methods of Disposal

Strongly acidic and alkaline materials, spent mordant baths, and dyebaths should be disposed of safely. This is not always an easy subject upon which to make recommendations.

Until recently, chemicals were categorized as nontoxic (nonpoisonous), toxic (poisonous), or highly toxic (highly poisonous). However, as our information and understanding has increased, along with lawsuits, the system has been revised, so that now all chemicals are labeled as being hazardous to varying degrees. For example, cream of tartar may carry the label "health hazard slight," potassium alum "handle with care," and sodium chloride (common table salt) "health hazard moderate" and "eye irritant."

Dyebaths

It would seem that most dyebaths derived from plants such as goldenrod, Queen Anne's lace, madder roots, etc., or cochineal would be classified as "health hazard slight." Microorganisms grow in these readily after a few days' standing. Cochineal is still used in cosmetics, and was our red food dye before synthetic red dye no. 2 or the synthetic acid dyes now in use.

My recommendation is that dyebaths produced from flowers, leaves, barks, plant stems, and insects, in practically all cases, be disposed of down the drain or on the ground away from the garden spot. These materials die and decay on the ground by the millions of tons each year, and their decomposition products enter the soil and water supply. The only exceptions which come to mind are dyebaths to which certain mordants have been added and dyebaths below pH 3 (where acid has been added). Dyebaths below pH 3 should be diluted and flushed with plenty of water if disposed of through concrete drain systems.

Any dyebath made acidic by addition of small amounts of oxalic acid should be disposed of in a hole in the ground and covered with earth. Selection of a spot to dig the hole is important. It should not be near septic tanks, wells, the garden, or places frequented by pets and children. Oxalic acid should always be used with care, and only in small amount.

Mordants, Acids, and Alkalis

The proper way to dispose of spent mordant baths depends upon the mordant and amount. Alum is the least toxic, followed by tannin, iron, tin, and copper, with chrome having the highest toxicity.

Spent alum baths may be disposed of down the drain with plenty of water, in the ground as previously described, or on your acid-loving plants, such as azaleas. Aluminum sulfate is sold in garden stores just for this purpose, and aluminum is the most abundant element in the earth's crust.

Iron and tannins, also abundant in nature, are usually used in relatively small amounts by home dyers. These spent mordant baths may be disposed of in the same way as alum, except that I do not recommend pouring them around acid-loving plants, even though they are slightly acidic and in small amounts would do no harm.

Tin and copper salts are less frequently used than alum, iron, and tannin, but they are also more toxic. Fortunately, they are used in small amounts. Again, the safest method of disposal is in the ground, away from gardens, wells, septic tanks, and where children and pets play. In small amounts (less than 1 ounce), and well diluted, these probably pose little, if any, environmental hazard if poured down the drain with plenty of water as well.

Chrome, potassium dichromate, is the most toxic of all the mordants. The spent bath should be buried in small amounts (less than 1 to 2 ounces) and never poured down the drain. Chrome dust, particularly, should not be inhaled, and the material should not be touched with the bare hands. Rubber gloves should always be worn when working with chrome.

I do not recommend pouring any mordant chemicals down the drain if they are to go into a septic tank. The proper balance in a septic tank is sometimes quite fragile even under the best of circumstances.

Acids stronger than pH 3, and alkalis stronger than pH 10 should be diluted with plenty of water during or prior to disposal. The pH may be measured accurately enough with purchased pH paper. To dilute acids or alkalis always pour them — slowly and carefully — into water, never the reverse. Also, remember that the pH scale is logarithmic so that a solution of pH 1 is 10 times as acidic as one of pH 2 and 100 times as acidic as pH 3. Vinegar, apple cider, and most soft drinks are pH 3 to 3.5.

At this time, compounds of lead should be disposed of only by a toxic waste disposal unit.

My recommendations for disposal are based on discussion of the problem with chemists and toxicologists, and on my knowledge of chemistry and ecology. A specific recommendation may not fit all areas of the country and all microhabitats. If you are at all in doubt, consult the toxic waste disposal unit in the area in which you live. When doing so, indicate the quantity of specific material to be disposed of since this may have considerable bearing on method of disposal. For example, certain toxic substances are permitted to be disposed of down the drain, but only if the amount is small enough, and only so much per day.

Scouring

Proper scouring is absolutely essential to good dyeing. Improperly scoured items do not dye level, the dye does not penetrate well, and the dyed item will likely never be rubfast. One reason why cotton appears to dye so poorly in comparison to wool is that cotton must be scoured by a different method than that used for wool. Woolen yarn, spun from unwashed fleece, is most difficult to scour well. I have found the following procedures successful for scouring the various fibers:

Cotton

1. Use a large enough nonreactive vessel so that the yarn or piece goods may be well covered and not crowded. The minimal amount of water is 1 quart per ounce of yarn or 2 quarts per ounce of piece goods. Cotton lint may be scoured and dyed, but it is most difficult to card and spin.

2. Fill the vessel with the required amount of water; add 1 teaspoon detergent per gallon of water and 2 teaspoons of washing soda per ounce of cotton. (Note: An inexpensive source of washing soda (soda ash, sal soda, Solvay soda, sodium carbonate) should be located. I obtain mine from Industrial Colloids here in Knoxville at a cost of 20 to 60 cents a pound. A 100-pound bag may cost as little as $20 to $30.)

3. Add the piece goods or skeined yarn and simmer or boil for an absolute minimum of 2 hours. Three or 4 hours is better in some cases, and in the old days cotton was sometimes scoured at the boil for 8 hours. Cotton is full of wax, pectic substances, and oil, all of which must be removed. The resulting wash water will be yellow-brown, and the odor will appeal to some, but not to others. (Note: Don't be fooled by nicely bleached, beautiful white cotton yarns. The chances are that these need a good scouring as well. Fabrics sold as "ready for dyeing," may be all right to dye without scouring.)

4. Rinse thoroughly 2 or 3 times.

5. Bleach with chlorine bleach only if very light colors are to be dyed, and if the material is not to be mordanted with tannin. Rinse very well following any bleaching.

6. Leave wet if the material is to be mordanted; otherwise dry and store.

Linen

Most linens cannot be treated with solutions as strongly alkaline as those used with cotton, but they also don't need as much alkalinity because linen fibers don't contain cotton seed oil. The following formula is from Elaine Gwynne (1982):

For 6 ounces of linen:

1. Add 3 tablespoons of Calgon water softener, 1 ounce of washing soda, and 1 ounce of Ivory Liquid to 2 gallons of room-temperature water. Stir until all components are dissolved.

2. Add the linen, heat to the simmer, and keep at that temperature for 1 hour.

3. Rinse well once and repeat the operation.

4. Finally, rinse well 2 or 3 times and mordant or dry and store.

Wool

1. Use 4 to 6 gallons of hot (120° to 140° F) water per pound of wool. Any nonreactive vessel is suitable.

2. Optional: add 1 tablespoon of Calgon water softener, particularly if the water is hard.

3. Add 1 to 2 level tablespoons Ivory Liquid, Tide, or similar detergent.

4. Place the yarn, fleece, or piece goods in the solution. If yarn or fleece, do not agitate. Agitation causes felting. The material should remain in the solution for at least 2 hours—overnight is excellent. Turn the material occasionally. If the yarn or fleece is especially dirty or oily, the material may be removed, gently drained, and placed in a new identically prepared solution.

5. Finally, rinse gently with water of approximately the same temperature as the scouring water at the time of removal. The material may be mordanted immediately or dried and stored. (Note: Fleece is best placed in a loose mesh bag such as those grapefruit or oranges are sold in. Dale Liles often scours fleece in a laundry basket, placed in a laundry tub. She scours the fleece in layers of locks, which preserves the arrangement found in the sheep, to reduce the chance of felting. She does not agitate fleece or pour water directly on it.)

Silk

Silk must be scoured (degummed) unless it is in the form of scarves, piece goods, etc., sold as "ready to dye."

1. Use 8 gallons of room-temperature water per pound of silk.
2. Add 2 to 3 tablespoons of Ivory Liquid detergent or 4 ounces of Ivory soap.
3. Heat up to the simmer and keep at that temperature until the silk no longer appears or feels slimy. This may take 30 to 60 minutes.
4. Remove, cool, rinse thoroughly, mordant or dry and store.

Mordanting

Unless substantive dyes are to be used, the next procedure that must be performed correctly is mordanting. The most frequently used method is premordanting (mordanting prior to dyeing). Occasionally, the mordant is added to the dyebath (one-pot dyeing), or the mordant is added near the end of the dyeing period (aftermordanting). Sometimes premordanting and aftermordanting are necessary or desirable.

Proper mordanting methods for wool are virtually never correct for cotton and linen, and often not for silk. Premordanting cotton with wool methods usually produces mediocre to poor results and is a waste of time, chemicals, and fiber.

It has been generally conceded that it is not necessary to use mordant chemicals rated as "reagent grade" or "chemically pure" and that "technical grade" is adequate. This is true unless the technical grade chemicals contain appreciable amounts of iron contamination, which will dull bright colors, especially reds and yellows. U.S.P. grade chemicals are sufficiently free of iron, however, and are less expensive than reagent grade.

Cotton and Linen Mordanting

In general, metal mordants do not combine chemically with cellulose fibers as readily as with the protein fibers. Fortunately, tannic acid (tannin) is readily absorbed by cellulose. Once mordanted with tannin, metal mordants combine well with the fiber-tannin complex. Oxyfatty acids developed from rancid olive, sesame seed, and castor oils (Turkey red oils) operate similarly to tannic acid. In addition, precipitated minerals such as iron oxide (rust) on cotton and linen form a good mordant for natural dyestuffs. Also, certain forms of mordants are more effective on cotton and linen. For example, "basic alum" and aluminum acetate are far more effective than the alum-tartar used on wool. In mordanting wool with alum, cream of tartar is added, which makes the solution more acidic, but in mordanting cotton with alum, washing soda is added, which makes the solution less acidic.

Tannin

For 1 pound:

1. Dissolve 1/2 ounce (light colors) to 1 1/2 ounces (dark colors) of tannic acid in 4 to 6 gallons of hot water (130° to 170° F) in a nonreactive vessel. If grey, black, or a dull color is desired, an iron vessel may be used.

2. Place the well-scoured fiber in the bath, work for a few minutes (remember the rubber gloves), then sink the material below the surface of the liquid and steep for 8 to 24 hours. Do not heat the bath again; in fact, cotton mordanting occurs well at room temperature.

3. Remove, rinse once, and dry or mordant with a metal mordant.

Note: Tannic acid is the best tannin source for bright and light colors. However, if tannic acid is not on hand, tannin-bearing plants such as sumac shoots and leaves may be boiled for 30 to 60 minutes to extract the tannin. For 1 ounce of tannic acid, use any of the following: 4 ounces dried or 8 ounces fresh sumac leaves and shoots, 2 ounces cutch extract or tara powder, 10 ounces oak galls, 18 ounces myrobalans, or 14 ounces divi-divi.

Basic Alum Mordant No. 1

This is the oldest basic alum mordant. It is quite satisfactory, though not as good as Basic Alum Mordant No. 2. It usually will not mordant well without tannic acid. For 1 pound:

1. Dissolve 8 ounces of potassium alum or aluminum sulfate in 2 to 3 quarts of hot water (130° to 170° F).

2. When the solution is cool, dissolve 1 ounce (5 level teaspoons) of washing soda in 1 pint of water. Add the soda solution slowly with stirring. There will be a large release of bubbles (carbon dioxide gas, as from a bottle of carbonated beverage). The mordant may be used as soon as the gas evolution ceases.

3. Dilute sufficiently with room temperature water to cover the material, but no more. This will amount to 3 to 5 gallons for one pound of cotton or linen. Alum cotton and linen mordants work best in concentrated form. The mordant may be used several times, but will be weaker with each use.

4. Add the well-scoured material, work for a few minutes, sink below the surface, and allow the material to remain in the mordant at room temperature for 6 to 24 hours. Six hours is sufficient for light colors or if the material is already tannin mordanted.

5. Remove and squeeze out. For best results, dry and air for one day before use. Rinse thoroughly two or three times before dyeing, or better yet, treat with fixing solution (Note C).

Note A: Piece goods should be dried by clothespinning the corners, preferably to a nonabsorbent plastic clothesline.

Note B: For best results, use this mordant twice following tannin mordanting, or before and after tannin mordanting.

Note C: Alum-mordanted cellulosics are best treated with "fixing solution," just prior to dyeing. Work and soak the material in hot (110° to 150° F) sodium phosphate (Na_2HPO_4) or powdered chalk ($CaCO_3$) solution for 30 minutes, then rinse well. The solution is made by adding 1/2 to 1 ounce of phosphate or chalk or 1 to 2 ounces of cattle or sheep dung to each gallon of hot water. The solution serves to fix the alum and remove unfixed alum. Dung contains sodium and calcium phosphates and was used for centuries for this purpose. Rinse thoroughly following use of the fixing solution.

Basic Alum Mordant No. 2

For 1 to 2 pounds:

This mordant is usually superior to Mordant No. 1, particularly if no tannin is to be used, but it is not as good as aluminum acetate.

1. Dissolve 1 pound of aluminum sulfate or potassium alum in 2 quarts of hot water (110° to 150° F). Allow the solution to cool to room temperature.

2. Dissolve 1 1/2 ounces of washing soda in one pint of room-temperature water. When it has dissolved, add this slowly, while stirring, to the aluminum sulfate solution. Bubbling will occur as carbon dioxide gas is released.

3. Make a paste of 1/2 ounce of chalk. Add this slowly, while stirring.

4. When all of the carbonic reaction ceases, add 1 ounce of 50% acetic acid or 1/2 ounce of 99% acid. Stir.

5. The mordant at this point is quite concentrated. Add sufficient water so that the material to be mordanted will be covered, but no more. For 1 pound of material, this may require a volume of 3 to 4 gallons.

6. Add the well-scoured fiber, work for a few minutes, sink below the surface, and leave the material in the mordant for 4 to 24 hours, time permitting.

7. Remove, squeeze out, and, time permitting, dry before use.

8. Just prior to dyeing, rinse thoroughly several times, or better yet, treat with "fixing solution" (Note C, Mordant 1).

Note A: Mordant twice for bright colors, or if the mordant volume exceeded 4 gallons.

Note B: The quantities listed are for 1 pound of cotton or linen; the mordant may be used for one or more additional batches, but these will be lighter in color.

Note C: This is an expensive mordant unless it is made with garden store grade aluminum sulfate. Aluminum sulfate is sold as soil treatment for acid-loving plants. In this form, the chemical may contain iron contamination. Iron contaminated aluminum sulfate is brown in color. If in doubt, dissolve an eighth of a teaspoonful of the aluminum sulfate in about an ounce of water. Add a few crystals of potassium ferrocyanide and stir. If the solution turns blue, iron contamination is present. The blue color results from Prussian blue formation, which is iron dependent.

Alum Mordant No. 3, Aluminum Acetate (Liles Method)

Aluminum acetate is the best alum mordant for cotton and linen, especially when used alone rather than in combination with tannic acid. Used alone, it produces good reds with madder or alizarin and good yellows with black oak bark, goldenrod, Queen Anne's lace, etc. It also produces good oranges with mixed red and yellow dyes. It is also an excellent alum mordant for silk, though no better than straight aluminum sulfate. Unfortunately, it is a rather expensive mordant, particularly if made from purchased calcium acetate which may be difficult to obtain.

The mordant came into use sometime between 1750 and 1790 and became the preferred alum mordant, particularly for calico printing, throughout the nineteenth century. Once made, it lasts about 4 weeks, and may be used several times, until exhausted. In former times, aluminum acetate often went under the names "acetite of alumina" or "red liquor."

My method of making aluminum acetate is the result of experimentation following chemical logic and following a suggestion made by Edward Bancroft in 1794. Bancroft suggested the use of calcium acetate instead of lead acetate for production of the mordant. The white sediment produced by this method, which must be disposed of, is the relatively nontoxic calcium sulfate (plaster of Paris) rather than the toxic lead sulfate. The recipe given will mordant 1 pound of cotton or linen, but the mordant may be reused for additional batches in lighter color intensities, until exhausted.

1. Dissolve 1 pound of potassium alum or between 2/3 and 1 pound of aluminum sulfate in about 3 quarts of hot water (110° to 150° F).

2. When it is cool, add gradually, while stirring, 1 ounce of powdered chalk made into a thick paste with water.

3. Dissolve 1/3 pound calcium acetate in 1 quart of water, and add it, while stirring, to the alum-chalk mixture. Stir for 2 to 3 minutes, and in about 30 minutes decant the clear.

4. Fill the vessel again with water, 3 to 4 quarts, stir again, and allow 30 minutes for settling. Decant again and repeat as before. Don't be concerned if the material contains some suspended white precipitate.

5. Bottle the mordant tightly in gallon jugs, or use immediately. When ready to use, add enough water to the mordant to cover the material, but no more than necessary. The mordant is used at room temperature.

6. Add the well-scoured cotton, linen, or silk, (wet or dry), and work well for about 5 minutes. Sink and allow the material to steep for an absolute minimum of 1 hour. Four to 6 hours is usually better.

7. Squeeze out, and dry slowly. Clothespin piece goods. Do not dye the material until the acetic acid smell has departed. This may take 1 to 4 days. One week was often allotted in former times. If extremely bright colors are desired, mordant again, using the same batch of mordant.

8. Treat with fixing solution (Note C, Alum Mordant 1) or rinse very thoroughly, just prior to dyeing.

Note: In former times lead acetate was usually used instead of calcium acetate. Twice as much lead acetate is required, by weight, because lead is much heavier than calcium. Because lead acetate is extremely poisonous and is suspected to cause cancer, its use is highly discouraged.

Alum Mordant No. 4, Aluminum Acetate

This method is modified from Molony (107) and is the only relatively inexpensive way of producing aluminum acetate from readily available materials. The disadvantage is that if the lime and aluminum sulfate are purchased at a garden store, they may contain some iron contamination. This recipe will mordant 2 to 5 pounds of cotton or linen.

1. Take 1 gallon of acetic acid, specific gravity 1.03, and add slowly and by degrees 2/3 to 1 pound of finely ground lime (calcium oxide). Use 1 pound if the lime is obtained from a garden store.

Note: Add 1 quart of concentrated (glacial) acetic acid slowly and carefully to 3 quarts of water, or 1.25 quarts of 80% acid plus 2.75 quarts of water, or 2.5 quarts of 40% acid plus 1.5 quarts of water. This produces acetic acid of the proper specific gravity (strength).

2. Stir well, occasionally, for 2 hours.

3. Let settle 48 hours. Then decant the clear carefully. If tested, the solution should read about specific gravity 1.09.

4. Add 2 quarts of warm water to the sediment. When it has settled, add the clear to the first clear. Now the sp. gr. will be about 1.06. Discard any remaining sediment. The clear is the calcium acetate.

5. To each gallon of the clear calcium acetate slowly add and dissolve 2.5 pounds of potassium alum or 2 pounds of aluminum sulfate. Stir well.

6. Let settle for 24 hours. The clear should stand at sp. gr. 1.09.

7. Decant the clear and add 2 quarts of water to the sediment. Let stand 24 hours.

8. Add the clear of this to the clear of 7. The sp. gr. should be about 1.06. The mordant may be used as is or diluted by not more than a half with water for yarn and piece goods. (Note: According to Molony, the undiluted (sp. gr. 1.06) product is good enough for printing red on the cylinder or yellow on the block, but if thickened with gum, must be sp. gr. 1.09. Here Molony is referring to the method of making calico printed fabric by printing the mordant by metal cylinder or carved wooden block, prior to dyeing.)

9. Be sure to treat mordanted material with fixing solution (see Note C, Alum Mordant No. 1) or rinse thoroughly before dyeing.

Copper Sulfate

Cellulosics may be premordanted with copper sulfate alone, as in dyeing logwood blue, or following tannin mordanting. Occasionally, it is used as an aftermordant, particularly with yellows dyed on alum mordanted materials. This treatment imparts a slight greenish cast to the yellow, but renders the color far more lightfast. Copper is added to the dyebath with cutch for browns. Recipe is for 1 pound.

1. Scour well, rinse, and leave wet.
2. Use a nonreactive vessel; iron is satisfactory for blue with logwood. Add 4 gallons of water to the vessel and heat to about 150° F.
3. Dissolve 3/4 to 1 1/4 ounces of copper sulfate in the hot water.
4. Add the yarn or piece goods, turn occasionally, and mordant for about 1 hour. Keep the temperature at scalding heat (150° F).
5. Remove the material, squeeze out, and dry or dye immediately.

Note: 1 1/4 ounces will produce navy blue with logwood if the copper sulfate and logwood are of high quality. One-half to 3/4 ounce is plenty when mordanting after tannin or as an aftermordant. One-half ounce is used with cutch for browns (one-pot method). Recipes for some dyes call for verdigris (copper acetate) instead of copper sulfate.

Iron-Iron Sulfates, Chlorides, Nitrates, and Acetates

Iron salts by themselves produce iron buff colors which vary from yellow to reddish orange. Iron salts with tannins produce colors ranging from slate or steel to grays and light blacks. Cellulosics take iron salts well; therefore, mordant dyes may follow, but the colors will be dark, dull, or sad. Some of these are quite beautiful if done with cochineal, and iron-alum combinations with madder produce Egyptian purple, violet, and brown. Indian dyers produced grays, blacks, and purples with iron-tannin-madder combinations. Iron is also used in the one-pot method and as an aftermordant with certain cotton and linen dyes. In former times, iron mordants were often in the form of "iron liquors," i.e., iron dissolved (reacted) in nitric, hydrochloric, and acetic acids. Appendix B contains directions for these mordants for dyers with proper chemical training, and for historical purposes. Recipe is for 1 pound.

1. Work and steep the well-scoured material 15 to 30 minutes in a bath prepared by completely dissolving 1 teaspoon to 2 tablespoons

of ferrous sulfate in 4 gallons of warm water. The greater the amount of iron used, the heavier will be the degree of mordanting, though heavier mordanting may be effected by repeating the process, using the same bath.

2. Remove, squeeze out and work and steep the material for about 15 minutes in a 4-gallon bath of warm washing soda solution. To make the solution dissolve approximately 4 level tablespoons of washing soda in the warm water.

3. Remove, squeeze out, and hang up to air. In 5 to 10 minutes the material will be finished. The entire procedure, starting with the iron solution, may be repeated for a heavier mordanting. At this point the material is mordanted (dyed) iron buff.

Note A: If tannin is to be used as well, mordant with the tannin first, and use only small amounts of iron salts (1 to 4 teaspoons). Additional iron may be added if deemed necessary. Iron tannates produce steel to gray colors.

Note B: Treatment with the soda bath is not always necessary, particularly if ferric salts are used.

Note C: If "iron liquors" are used, a small amount (1/2 to 1 ounce) is added to the bath initially, and more added later if deemed necessary. Alkaline (washing soda) baths are generally not used with iron liquors.

Note D: Iron mordanting can be tricky. It is best to keep turning the material to avoid streaking and spotting. If spotting occurs, place the material back in the iron bath and work again for several minutes.

Chrome

On cellulose fibers, chrome is mostly used as an aftermordant, as in logwood blacks, and as an oxidizing agent with cutch. Chrome-alum is used as a mineral dye in "sea green." Use of chrome is described in the recipes requiring it.

Tin (Stannic and Stannous Chlorides—"Red Spirits")

Tin mordants were used rather extensively, by themselves, particularly with redwood dyes and logwood purples. These dyes are quite pretty, newly made, but they are neither lightfast nor particularly washfast. The colors do not fade true; instead, a rather ugly off-brown is the usual final result. Tin mordants were also used as brighteners, but this practice is discouraged because of the fading problem. Directions for production of "red spirits" are given in Appendix B. Tin mordant produced from stannic chloride follows:

1. Dissolve 3 ounces of stannic chloride in 1 gallon of soft or deionized water. This is the mordant.

2. Add no more than 4 ounces of well-scoured cotton or linen, work well for a few minutes, sink and steep for approximately 1 hour. After 1 hour, the material should be a light lemon color.

3. Remove, rinse well, and leave damp.

4. Mordant the remainder of the material in similar fashion (up to 1 to 1 1/2 pounds).

5. The material is now ready for the dyepot.

Turkey Red Oils

Directions for production of these mordants are listed with the Turkey red recipes (see Red chapter). Oil mordants were occasionally used for other colors, such as madder purple.

Silk Mordanting
Alum

For 1 pound of silk:

1. Dissolve 8 to 16 ounces of aluminum sulfate or potassium alum in 3 gallons of hot water. Use a nonreactive vessel. When cool, the mordant is ready to use.

2. Add the well-scoured wetted-out silk. Work well for several minutes, sink and steep for several hours.

3. Remove, carefully squeeze out, and hang up the silk. When it is nearly dry, repeat the process if very deep colors are desired.

4. Rinse well and add to the dyepot.

Note A: The mordant may be used several times, until exhausted.

Note B: Some recipes from the old literature call for as much as 1 pound of alum per gallon of water.

Note C: Wool alum-tartar mordant may be substituted, but in that case the mordanting must be done hot, which is destructive to the luster of the silk.

Note D: Use of a garden-store grade of aluminum sulfate is the only inexpensive way to make this mordant.

Iron (Ferrous Sulfate and "Black Iron Liquors")

Iron mordants are used on silk, usually at the rate of 1/3 to 1 ounce of ferrous sulfate per pound of silk, for grays and blacks. The best iron mordant for silk is probably ferric chloride (Appendix B). Directions for use of iron vary and are listed with individual dye recipes.

Tin (Stannous Chloride—"Red Spirits")

On silk, tin is occasionally used as a premordant, as a one-pot mordant, and as an aftermordant. It is used primarily in red recipes and often in conjunction with other mordants. Directions for use of tin are given with individual dye recipes.

Tannin

Tannin mordanting for silk follows exactly the same procedure as that used for cotton (see Cotton and Linen Mordanting). As a rule, most recipes call for 1/2 to 1 ounce of tannin per pound of silk. When necessary, tannin is combined with alum, tin, or iron, and sometimes a combination such as alum and tin. Directions for these combinations are given in the individual dye recipes.

Wool Mordanting

Wool mordants are applied several different ways. Most commonly, the yarn, fleece, or piece goods are premordanted and then dyed. Occasionally, the mordant is added to the dyebath just prior to, or a little after, the material has been added (one-pot method). At times, a mordant (usually a second mordant) is added near the end of the dyeing cycle as a brightener or modifier. The following directions are for premordanting 1 pound of wool.

Alum-Tartar

1. Dissolve 3 ounces of potassium alum (86 grams or 5 1/2 level tablespoons granular alum) in 4 to 6 gallons of warm water. Use a nonreactive vessel.

2. When the alum is dissolved, add 1 ounce (29 grams or 10 level teaspoons) of cream of tartar. Stir until dissolved.

3. Add the well-scoured wet fiber and heat slowly to the simmer (about 190° to 200° F). Keep the material at this temperature for at least 1 hour, preferably 1 1/2 to 2 hours. Turn the material occasionally during mordanting. Allow the material to cool in the mordant bath, time permitting.

4. Remove fiber and squeeze out excess moisture. Dry without rinsing if dyeing is not to be done immediately. Slow drying, done in a cotton or linen bag, is best. This way the material remains damp for several days. Mordanted material may be stored.

5. Rinse very thoroughly before dyeing to remove unfixed alum which will loosely attach to the fiber and dye, producing a dye job which may not be rubfast. Items not well rinsed may also dye unevenly. This is true of all mordants.

Note A: Very fine wools often require only 2 ounces of alum, and 4 ounces of alum is sometimes required for very deep colors. Use of more than 4 ounces makes wool sticky.

Note B: Aluminum sulfate may be substituted for potassium alum. Use 2 ounces of aluminum sulfate instead of 3 ounces of potassium alum.

Chrome

1. Dissolve 3/8 ounce (2 level teaspoons) of potassium dichromate in 4 to 6 gallons of warm water in a nonreactive vessel.

2. Add the well-scoured wetted-out wool and a loose cover, or have the fiber well covered with liquid (chrome is apparently light-sensitive under certain circumstances).

3. Heat to the simmer and keep at that temperature until the color of the fiber and dyebath changes from orange to gray (usually 1 to 1 1/2 hours).

4. Remove the fiber, cool, and squeeze out.

5. Rinse well and dry for storage, or dye immediately.

Note A: Remember that chrome is the most toxic of the metal mordants and must be disposed of properly.

Note B: Some add 1/4 ounce of cream of tartar or tartaric acid or concentrated sulfuric acid to the chrome bath, but I have not found this to be necessary.

Tin-Tartar–Oxalic Acid

1. Dissolve 1 1/2 ounces of cream of tartar in 4 to 6 gallons warm water in a nonreactive vessel. When dissolved, add 1/2 ounce of oxalic acid. Stir until dissolved.

2. Add 3/8 ounce of tin (stannous chloride). Stir until dissolved.

3. Add the well-scoured wet fiber and heat slowly to the simmer. Keep the material at this temperature for 1 hour. Turn the material occasionally.

4. Remove, squeeze out, cool, rinse well, and dry for storage or dye immediately.

Note A: This premordant produces the brightest colors, but they are rarely very lightfast. Cochineal reds and scarlets are an exception.

Note B: Tin is sometimes used as an aftermordant to brighten dyed items premordanted with alum-tartar. To do this, dissolve 3/8 ounce of tin in approximately one quart of the hot dyebath liquor, lift the wool out of the bath, stir in the dissolved tin, return the wool, and dye for an additional 10–15 minutes. Remove the wool, cool, and rinse well before drying.

Note C: Never use excessive amounts of tin since it tends to make woolen fibers brittle.

Note D: In some cases the oxalic acid may be omitted. This is especially true of cochineal scarlets.

Copper-Tartar

1. Dissolve 3/8 ounce of cream of tartar in 4 to 6 gallons of warm water in a nonreactive vessel. When dissolved, add 3/4 ounce of copper sulfate. Stir until dissolved.

2. Add the well-scoured wet fiber and heat slowly to the simmer. Keep the material at this temperature for 1 hour. Turn the material occasionally.

3. Remove, squeeze out, cool, and rinse well.

4. Dry for storage or dye immediately.

Note A: This mordant, by itself, dyes the wool a nice permanent mineral sea green if high-quality copper sulfate is used.

Note B: 3/8 ounce of copper sulfate added to alum-tartar mordant will produce more lightfast yellows than alum-tartar alone but has a slight greening effect. This is usually not a disadvantage when overdyeing indigo blue with a yellow dye for green.

Iron-Oxalic Acid

1. Dissolve 1/2 ounce of ferrous sulfate in 4 to 6 gallons of warm water. When dissolved, add 1/2 ounce of oxalic acid. Stir until dissolved.

2. Add the well-scoured wet fiber and heat slowly to the simmer. Keep at this temperature for 1 hour. Move the material frequently.

3. Remove, squeeze out, rinse well, and dry for storage, or dye immediately.

Note A: Iron is rarely used as a premordant on wool and is difficult to mordant evenly. More often, iron is used as aftermordant to dull or darken a color. To do this, dissolve 1/4 ounce of ferrous sulfate in approximately 1 quart of hot dyebath liquor, lift the wool out of the dyebath, stir in the dissolved iron, return the wool, and dye for an additional 10 or 15 minutes. Remove the wool, cool, and wash thoroughly before drying.

Note B: Clean the dyeing and mordanting vessels very thoroughly following use of iron. Only a small amount of residual iron can dull subsequently dyed bright colors.

Alum-Copper-Tartar and Alum-Chrome-Tartar

1. Dissolve 1 1/2 ounces alum in 4 to 6 gallons of warm water. Use a nonreactive vessel.

2. When the alum is dissolved, add 3/16 ounce of chrome *or* 3/8 ounce of copper sulfate. Stir until dissolved. Finally, add 1 ounce cream of tartar. Stir until dissolved.

3. Add the well-scoured wet fiber and heat slowly to the simmer. Keep at the simmer for 1 1/2 hours.

4. Remove, squeeze out, cool, rinse well, and dry for storage, or dye immediately.

Note: These two compound premordants produce some very nice, lightfast colors. With a yellow dye, the color is usually between that produced by the alum and the chrome or the alum and the copper. With cochineal, nice purples result from using either of these compound mordants.

Bancroft's Mordant

1. This is a "one-pot" method. Therefore, dissolve 1/3 ounce of alum (2 level teaspoons) and 1/5 ounce of cream of tartar (2 level teaspoons) in the prepared, cool, 4 to 6 gallon dyebath. Stir until dissolved.

2. Add 1/3 ounce of tin (2 level teaspoons). Stir until dissolved.

3. Add the well-scoured wet wool immediately, and slowly raise the temperature to 140° to 160° F for yellows, and to the simmer for other colors.

4. Dye for 1 hour or less.

5. Remove, squeeze out, rinse well, and dry.

Note A: This method is very good for some clear yellows, especially black oak bark, and some reds using brazilwood, cochineal, or madder. Because this one-pot method involves fewer steps and requires only very small amounts of mordant chemicals, it is also of value in judging dye potential of unknown plant material. Another advantage of this mordant is in dyeing wool fleece for spinning, because in the one-pot method the mordanting and dyeing occur simultaneously, thus the material has to be heated to high temperature only once. This reduces felting of the fleece.

Note B: This method works well with only very small amounts of mordant chemicals. Therefore, lightfast qualities with most yellow dyes may only be fair. Addition of 1/4 the prescribed premordant quantity of copper or chrome 15 minutes before the end of the dyeing session will improve lightfastness, but copper has a greening effect and chrome an oranging effect.

Note C: Slightly stronger dyebaths are required with one-pot methods because some of the dye complexes with the mordant, but not with the fiber, and is lost.

Supplies and Suppliers

Listing suppliers can be risky since some are not in business for an extended period. Most advertise fairly frequently in fiber magazines. This includes suppliers of dyestuffs, yarns, fleece, and piece goods. A good listing can generally be obtained by making a library check of several issues of three or four appropriate craft magazines. Several establishments in business for a number of years are:

1. Mr. Roger Wolfe
   Alliance Import Company
   1021 "R" Street
   Sacramento, CA 95814
   (916) 920-8658
   Natural dyes (indigo, madder, cochineal, cutch, etc.), mordants (wholesale)
2. Cerulean Blue Ltd.
   P.O. Box 21168
   Seattle, WA 98111-3168
   (206) 443- 7744
   Indigo, cotton piece goods, chemicals
3. Fisher Scientific
   2775 Pacific Drive
   P.O. Box 4829
   Norcross, GA 30091
   (404) 449-5050
   Chemicals, alizarin, etc.

4. The Earth Guild
   Dept. HW
   1 Tingle Alley
   Asheville, NC 28801
   Natural dyes, mordants, yarns
5. Test Fabrics, Inc.
   P.O. Drawer O
   Middlesex, NJ 08846
   (201) 469-6446
   Cotton, linen, wool, and silk piece goods
   which are scoured, free of sizing, and ready to dye
6. Thai Silks
   252 State Street
   Los Altos, CA 94022
   (415) 948-8611
   Silk scarves and piece goods, ready to dye
7. Rumpelstiltskin
   1021 "R" Street
   Sacramento, CA 95814
   Imported natural dyes; mordants
8. The Unicorn
   Box 645
   Rockeville, MD 20851
   (301) 933-5497
   Fine quality Perle cotton yarns; books
9. Sigma Chemical Company
   P.O. Box 14508
   St. Louis, MO 63178
   (800) 325-3010
   Chemicals, alizarin, tannin, etc.

Larger cities usually have a chemical supply outlet which may well carry potassium alum (swimming pool alum), soda ash, and aluminum sulfate. Aluminum sulfate, sold as soil treatment for acid-loving plants, and copper sulfate are sold by garden stores. Farmers Coop stores generally carry lime (calcium oxide), copper sulfate, and ferrous sulphate (copperas). Chalk (calcium carbonate), potassium permanganate, potassium ferrocyanide, etc., may be obtained from a druggist or druggist supply outlet. Refer to Appendix A concerning outmoded chemical names.

3 Yellow Dyes

Yellow dyes are of extremely ancient use on all of the natural fibers, perhaps because of their accessibility. Indeed, hundreds of species of plants and some lichens yield yellow, gold, yellow-green, yellow-orange, or golden-tan dyes.

Unfortunately, most natural yellow dyes are more or less fugitive to light. It is for this reason that many old textiles show only weak yellows, oranges, and greens. The majority of the yellows, chemically, are flavins and flavones, which are adversely affected by ultraviolet light. However, there are a few notable exceptions. From Roman times to about 1800, the best-known fast yellow dye was weld (*Reseda luteola*). About 1775, Dr. Edward Bancroft discovered a highly concentrated yellow dye in the inner bark of the American black oak tree (*Quercus velutina*). This was an extremely significant discovery since this dye, which he named "quercitron," was as fast or faster than weld, and it was much cheaper because the dye in weld is present in lower concentration. Thus, quercitron was to become the best natural yellow dye for the next century and was used commercially until about 1920 (Matthews, 500).

Two yellow dyes used quite early in Europe were *Rhus cotinus* which was known as "young fustic," "Venetian sumac" or "fustet," and *Rhamnus* sp., called "Persian berries." Young fustic was particularly fugitive to light and was usually used only to heighten the shade of a more permanent, less brilliant yellow. Persian berries was fairly lightfast but not as good as weld.

Turmeric (*Curcuma longa*), or "Indian saffron," was used from earliest times, at least in India and China, for bright clear yellows and greens, particularly on chintzes and palempores. Thickened with gum, it could be painted on directly because it dyes substantively. Unfortunately, turmeric survives little ultraviolet and therefore cannot be recommended for anything subjected to much sunlight. It does last better in combination with other dyes for greens, olive-greens, and browns (Adrosko, 37) and was commonly used with cochineal in production of redcoat scarlets.

Chrome yellow (lead chromate), a mineral dye, became available by about 1840 and by 1870 was the most widely used clear yellow on cotton and linen. It is completely lightfast and retains its brilliant color very well, unless subjected to high concentrations of coal smoke (sulfur dioxide), which turns it gray. The dye was used as a yellow and as an overdye with indigo or Prussian blue for greens on cotton as late as 1920. The disadvantage of chrome yellow is that it is poisonous. This was not a problem for wearers of clothing dyed with it but was a considerable problem to factory workers producing large yardages of dyed cloth, many of whom must have suffered lead poisoning.

The yellow dye used commercially more than any other from approximately 1600 to 1850 was "old fustic." This dye, extracted from a small tree of the mulberry family (*Morus tinctoria*), grows indigenously in Brazil and the West Indies. It was brought back to Europe by the Spanish explorers (along with cochineal, logwood, and brazilwoods). Old fustic is not as lightfast as weld or black oak bark, but it was and is reasonably satisfactory, particularly when mordanted with alum-copper, copper, or chrome. Generally speaking, old fustic, Osage orange, Queen Anne's lace, and goldenrod are among the most satisfactory yellows for use with indigo in producing a variety of greens.

Iron buff, a mineral dye carefully made with pure ferrous sulfate, can produce a rather pleasing though slightly dull yellow color. As commonly produced, iron buffs are yellowish or reddish orange. Though dull, they were dyed on cellulose fibers from earliest times up to the beginning of the twentieth century because of their excellent lightfast and washfast qualities.

Saffron (*Crocus sativus*) dyes a very clear yellow with a slight orange tinge. Like turmeric, saffron dyes substantively but is fugitive to light. Even though extremely expensive, the dye was used rather extensively quite early in the Middle East, Egypt, and other Mediterranean countries, and in Europe during the Middle Ages. At current prices, about fifty dollars' worth of saffron would be required to dye a pound of fiber. The dye is present only in the stigmas of the female part of the flower.

Species of plants yielding good yellow dyes in the general area of southeast North America include black-eyed Susans, black oak bark, broom-sedge, white aster, coreopsis, goldenrod, mimosa leaves, marigolds, onion skins, Osage orange, peach leaves, Queen Anne's lace, smartweed, and tickseed or bidens. The most lightfast of these are black oak bark (*Quercus velutina*), Osage orange (*Maclura pomifera*), peach leaves (*Prunus persica*), Queen Anne's lace (*Daucus carota*), smartweed or arsemart (*Polygonum persicaria*), and goldenrod (*Solidago* sp.). Chemically, Osage orange is very similar to old fustic. All of these dyes are more lightfast if mordanted with copper or chrome. Copper has a greening effect and chrome an oranging effect. All do fairly well with alum, and a little added copper helps and, in small amounts, does not alter the color. Black oak bark is fast with alum alone, but its bright, clear yellow is not altered if the item is mordanted or dyed in a copper pot, or if a little copper is added.

The following recipes are for dyeing 1 pound of fiber.

Black Oak Bark *(Quercitron)*: For All Natural Fibers

1. Enlist the assistance of a botanist in locating a black oak tree, and, if convenient, read Gerber's article, "Quercitron, The Forgotten Dyestuff" (1978b).

2. Remove the bark from trunk or limbs with a tool such as an axe or knife, remembering that the dye is in the inner bark. Remove and discard the outer bark, and cut the inner bark into pieces. I prefer the pieces to be 2 to 6 inches long, 1 to 3 inches wide, and about 1/4 inch thick, but the bark from even the smallest branches works. Spread the cut-up bark out to dry (for storage). The bark, when dry, is fairly easy to shred or cut up fine.

3. Scour yarn or piece goods, rinse, squeeze out, and leave damp.

4. Premordant once or twice with aluminum acetate or Basic Alum Mordant No. 2 for cotton or linen, aluminum sulfate or acetate for silk, but only once with alum-tartar for wool. Work piece goods well to ensure even mordanting.

5. Rinse very thoroughly. Use fixing solution prior to rinsing with cotton or linen if it is on hand.

6. Prepare the dye liquor: Shred up fine 1 to 2 ounces of dry bark in 2 to 4 quarts of hot tap water. Use a nonreactive, copper, or brass vessel. Heat to 170° to 190° F and keep at that temperature for 1 hour or place a lid on the vessel once 180° to 190° F is reached and leave, preferably, overnight or until cool.

7. For 1 pound of cotton, linen, or silk, place 5 to 12 gallons of hot (120° to 140° F) tap water into a nonreactive, brass, or copper vessel. For wool, use the same quantity of water (5 to 12 gallons),

but at 140° to 180° F. Add half of the dye liquor, stir, and slowly enter the damp yarn or piece goods.

8. Work well for the first few minutes, since the dye will be taken up rather rapidly and working the material helps to ensure level dyeing.

9. Within 10 to 20 minutes cotton, linen, or silk may be dyed as dark as desired; wool usually takes longer. Remember that cotton and linen appear at least two shades darker wet than dry. If dark enough, lift out, squeeze out, and air the material temporarily. Add about 1/2 teaspoon of well-powdered chalk and stir well. Reenter the goods and work for several minutes longer. If not dark enough after the initial 20 minutes, add additional dye liquor and work for another 10 to 20 minutes before adding the chalk.

10. Remove from the dyebath, squeeze out, and rinse twice before hanging up to dry.

• Dyeing with this yellow, or any other vegetable yellow, at low temperature gives a clearer color. If a greenish yellow or gold is desired, then heat the dyebath up to near the boil or to boiling. Other off colors and tannins will be absorbed. In fact, in dyeing any vegetable yellow, remove the material from the dyebath when the color is right. Prolonged heating, particularly boiling, ruins most yellows.

• Using black oak bark with this recipe in a vessel which can be heated up to 180° to 190° F gives one of the very best old golds on cotton and linen.

• Quercitron is a very lightfast dye when mordanted with the alum mordants alone. However, the clear yellow color can still be retained, and the lightfastness even further improved, by dyeing in a copper or brass vessel or by adding a few clean pennies or clean brass cartridge cases or a pinch of copper sulfate or verdigris to the dyebath.

• With cotton, linen, or silk, if the dyed item turns out uneven or a darker yellow than desired, place the material back into the alum mordant, work, and leave submerged for an hour or two. This often helps to level and remove some of the color. The material can then be washed well or redyed.

• Black oak bark dyes beautifully on wool by the one-pot method worked out by Bancroft. A modernized "Bancroft Mordant" was produced through investigation by Fred Gerber and is described in the mordants section. This method is easier and produces a richer wool yellow than alum-tartar premordanting.

• Boiling black oak bark to extract the quercitron does not seem to affect the yellow as adversely as does boiling certain other barks.

- Chemically, quercitron consists of quercitron, $C_{36}H_{38}O_{20}$, and quercetin, $C_{22}H_{24}O_{11}$ (Matthews, 500).
- This basic recipe generally works well on cotton, linen, wool, and silk with all of the good yellow dyes except that more dyestuff will be required, and the addition of chalk is omitted. On cotton and linen, Aluminum Acetate Mordant produces the most washfast result, followed by Basic Alum Mordant No. 2. Basic Alum Mordant No. 1 should be alum-tannin-alum or tannin-alum-alum or alum-copper. Alum-copper mordanting is described below in "Old Fustic or Osage Orange: For Cotton and Linen."

Old Fustic or Osage Orange: For Cotton and Linen

This recipe is a modification of that of Bronson (122) and is typical of many vegetable yellow cellulose dyes of the eighteenth century. Use of copper sulfate as well as basic alum usually produces a reasonably lightfast and washfast result, though the color is usually greenish yellow.

1. Dissolve 4 ounces of potassium alum in 4 to 6 gallons of hot tap water in a nonreactive, brass, copper, or plastic vessel.

2. Dissolve 3/8 ounce (2 1/2 to 3 level teaspoons) of washing soda or pearlash in 1 quart of hot water. Add this solution slowly to the alum with vigorous stirring. When the carbonic reaction ceases, add the well-scoured cotton or linen and work well for several minutes. Remove, squeeze out, shake, and then sink in the mordant. Allow the material to remain in the mordant for 6 to 12 hours.

3. Remove from the mordant, squeeze out well, and hang up carefully. Piece goods should be hung up by clothespinning.

4. Prepare the dyebath by dissolving 1/2 ounce of fustic or Osage orange extract in 4 to 6 gallons of hot water. If chips or sawdust are being used, rather than extract, simmer 8 to 12 ounces of the material for 2 hours in 4 to 6 gallons of water. Pour the solution into another vessel and allow it to cool to a temperature of 130° to 160° F before dyeing.

5. Add the damp alum-mordanted material to the dyebath and work well. If the mordanted item has dried, rinse before immersing it in the dyebath. Work or move the material in the dyebath for about 30 minutes, then remove, squeeze out, and air.

6. Dissolve 1/2 ounce (3 1/2 level teaspoons) of copper sulfate in about 1 quart of the warm dyeliquor and add this to the dyebath. Make certain that the dyebath temperature has not dropped below

130° F at the time of addition of the copper. Reintroduce the cotton or linen and work well for about 10 minutes.

7. Remove, squeeze out, rinse, and dry.

• The recipe works well with black oak bark, peach leaves, goldenrod, and smartweed. It may work well with any good yellow dye. If using fresh plant material, crowd as much into the dyepot as it will hold of goldenrod, smartweed, or peach leaves. Use half that much if the material is dry. All of these should soak overnight (3 or 4 days with smartweed), then simmer for at least 2 hours. The smartweed (arsemart) may smell a bit.

Vegetable Yellows, General: All Natural Fibers

As far as my experiments have gone, most good plant-derived adjective yellow dyes work well on cotton and linen using aluminum acetate, basic alum, tannin-alum-alum, or alum-copper mordants as described in the black oak bark and old fustic–Osage orange recipes. Aluminum sulfate is recommended for silk, and alum-tartar for wool. With wool premordanted with alum-tartar, tin may serve as a brightening afterbath; iron as an afterbath dulls or darkens the color. Following are methods of collection and dyebath preparation for the major yellow vegetable dyes.

Broom-sedge (*Andropogon virginicus*):

Collect from poorly fertilized pastures at any time of the year. Cut the broom-sedge off close to the ground; use fresh, or it may be dried and stored. The shade of yellow will vary slightly depending upon the month of collection. For 1 pound of material, crowd as much of the dry broom-sedge as possible into a nonreactive 4-gallon vessel with enough water to cover. Soak overnight and cook (simmer or boil) for about 2 hours. Pour off the dyeliquor, which is ready to use. Heat it further to concentrate the dyeliquor for storage. Freeze to store; or add .1% sodium benzoate (about 1 teaspoon/gallon).

Goldenrod (*Solidago sp.*) and Queen Anne's lace (*Daucus carota*):

Collect goldenrod when the flower heads are just fully open or nearly so, and Queen Anne's lace when it first comes into full bloom. The plants collected at this time produce the clearest yellows. Queen Anne's lace may be cut off close to the ground, but use just the goldenrod flower heads. It will take about 1 pound of fresh Queen Anne's lace or 1/2 pound of goldenrod per pound of fiber. These two plants do not dry well for storage and should be cooked up soon after picking. Treat Queen Anne's lace the same as broom-sedge, but goldenrod may be cooked immediately without soaking. Since goldenrod is more concentrated, it is a good dye to prepare, place in plastic jugs, and freeze for later use.

Smartweed *(Polygonum persicaria)*, peach leaves *(Prunus persica)*, and mimosa leaves *(Albizia julibrissin)*:

All of these may be collected, dried, and stored. Collect the smart weed when the little pink flowers are in bloom. Smartweed is often found growing in damp places, including along interstate highways. Cut the plant close to the ground, as the dye is present in all parts of the plant. Strip mimosa leaves from the tree branches when the pink-yellow flowers are in bloom. It is at this time that the dye is best. Peach leaves should be stripped from the branches before they are damaged too much by insects. One pound of dried smartweed or peach leaves will be required for each pound of fiber to be dyed. For mimosa leaves, 2 to 4 ounces of dried leaves will suffice. Soak peach or mimosa leaves overnight and then cook. Smartweed should be soaked 3–4 days before cooking. Use or store the dyebath as for broom-sedge or goldenrod.

Black-Eyed Susans *(Rudbeckia serotina)*, aster *(Aster pilosus)*, coreopsis *(Coreopsis* sp.), tickseed *(Bidens* sp.), and marigolds *(Tagetes* sp.):

All of these flowers should be picked when fully open (the entire aster plant may be used). All may be dried or frozen and stored. In general, it will take 1 gallon of fresh flowers or 1/2 gallon dried flowers for each pound of fiber. Prepare the dyebath as for goldenrod (presoaking prior to cooking is not essential).

Osage orange *(Maclura pomifera)*:

This dye is in the wood of all parts of the tree, but not the bark. Cut the wood repeatedly, saving the sawdust, or have the wood chipped into small pieces. The wood contains a lot of dye. The chips or sawdust should be simmered for 2 hours to extract the dye. Eight to twelve ounces of chips or sawdust will produce a bright yellow with 1 pound of fiber.

Dyeing Wool Yellow, Using "Bancroft's Mordant"

Dale Liles and I have been using Bancroft's Mordant to dye wool yellow for a number of years. Some items have been left yellow and others overdyed for oranges and greens. Black oak bark yellows are particularly clear and bright, and produce excellent greens overdyed with indigo or indigo sulfate. We have also gotten good results with wild white aster, chrysanthemums, bidens, yellow onion skins, marigolds, Osage orange, old fustic, mimosa leaves, and southern red oak bark.

Eventually, we published an article on our results in *Shuttle, Spindle and Dyepot*. The pictures of the dyed yarn, in color, are quite accurate in most cases. The article was also published in *Dyeing for Fibres and Fabrics* (the Australian Forum for Textile Arts), but without the color swatches.

We are not certain whether lightfast qualities are as good as with premordant methods, particularly since no copper or chrome is used, but the method is certainly an easy, quick, and convenient one-pot procedure. Thus, the primary advantages of the method are time saved, economy of chemicals, brightness and clarity of color, and, with fleece, a decreased tendency to felt because the wool is heated to relatively high temperature only once.

Initially, we used Bancroft's Mordant to exhaust the color in dyebaths previously used with premordanted yarn. We found that light but bright colors often result, but if not, the Bancroft-mordanted yarn or fleece is then a premordanted, partially dyed product that may be overdyed in another exhaust bath without additional mordant, or used with a weak indigo, madder, cochineal, brazilwood, or combination red overdye. This method is also quite useful in determining the dye potential of an unknown plant or of old, stored dyestuffs.

1. Scour the wool, rinse, and leave wet.

2. Place 4 to 6 gallons of prepared yellow dyebath in a nonreactive, brass, or copper vessel.

3. Dissolve 1/3 ounce (2 1/2 to 3 teaspoons) of potassium alum in the room-temperature dyebath. To this solution add 1/5 ounce (1 1/2 to 2 teaspoons) cream of tartar (omit this with black oak bark). Stir until dissolved.

4. Dissolve 1/3 ounce (1 1/2 teaspoons) of tin (stannous chloride) in a quart of hot water. Add this solution to the dyebath and stir.

5. Add the wet yarn, fleece, or piece goods and move it about in the dyebath.

6. Heat the dyebath slowly up to 160° to 180° F, and keep it at that temperature for 15 to 45 minutes. Lift and air the material 2 or 3 times during the process. The oxidation in the air seems to intensify the dyeing potential. This was often done in the old days, and is highly recommended by Gerber. We agree.

7. Remove the wool when the color looks right. Overcooking yellows produces dingy colors.

8. Remove, squeeze out, cool, rinse well, and dry.

• This method also works well with madder, cochineal, brazilwood or a combination of these reds.

Turmeric: Cotton, Linen, Wool, and Silk

Remember that turmeric (curcumin) is fairly fugitive but can be redyed. It is also affected by alkalis which turn it red-brown. Newly made, it is a bright clear yellow. Since turmeric dyes substantively and is used in cooking, it is an excellent dye for children's projects.

1. Obtain the turmeric from the grocery store.

2. Dissolve 1 1/2 to 2 ounces of turmeric in 4 to 6 gallons of water in a nonreactive vessel.

3. Add the well-scoured, damp material.

4. Heat the vessel slowly to 160° to 190° F. Dye at this temperature for about 30 minutes.

5. Remove, squeeze out, cool, rinse well, and dry.

Curcumin has the formula $C_{21}H_{20}O_6$ and is a diferuloylmethane (Mayer–Cook, 93).

Saffron: Cotton, Linen, Wool, and Silk

Saffron, like turmeric, is fugitive but can be redyed. At present, saffron is used only to color and flavor food. I obtain mine from the Indian market here in Knoxville. The dried flower stigmas (whole) are sold in 1-gram quantities, at three dollars a box. This is enough to dye a 1-ounce scarf.

1. Prepare the dyebath by placing 1 gram of saffron in 2 gallons of water. Use a nonreactive vessel. Simmer until all of the color is gone from the stigmas.

2. Cool the dyebath to about 130° to 140° F, then add the well-scoured, damp material. Dye at 130° to 150° F for 20 to 30 minutes.

3. Remove, squeeze out, cool, rinse, and dry.

The dye in saffron is a glycoside called crocin, $C_{44}H_{64}O_{26}$ (Mayer–Cook, 71).

Chrome Yellow: Cotton and Linen

Caution: *remember that chrome yellow is poisonous.* Chrome yellow may be produced from the palest shade to deep yellow, and it is extremely lightfast. This was the main commercial cotton yellow dye from about 1850 to 1910 (Matthews, 513). It was also used to color paper, and our "greenback dollars" were dyed with chrome green (Prussian blue overdyed with chrome yellow) from about 1850 to 1900 (Pellew, 94). Chrome yellow can be discharged, as is the case with all mineral dyes, and so was used in calico printing.

1. Light, medium, and heavy shades of yellow will require respectively 1, 2, or 3 ounces of lead nitrate or acetate, and 1/3, 2/3, or 1 ounce of potassium dichromate.

2. Dissolve the lead and chrome in 3 to 4 gallons each of room-temperature water in separate plastic or nonreactive pails.

3. Work the well-scoured damp or dry cotton or linen in the lead nitrate bath for about 15 minutes (better penetration occurs if the cotton is dried before placing it in the lead nitrate).

4. Remove from the lead solution, squeeze very thoroughly, and work for a few minutes in the chrome. The yellow color will develop almost instantly, but the material should remain in the chrome for at least 10 minutes. It is important to move the material to ensure level dyeing.

5. Remove, squeeze out, and check for depth of color and level dyeing. If it is not level, or deep enough, repeat the procedure, using the same baths. If the color is deep enough and dyeing is level, place back in the lead solution only, and work for 5 to 10 minutes.

6. Remove, squeeze out, and wash thoroughly with detergent. Rinse several times. This removes surface dye which could dust off. Hang to dry.

• If using this dye to produce green, remember that the Prussian blue or indigo *must* be dyed first.

• Chrome yellow is not suitable for wool or silk, and since it is poisonous should be made only in small quantities out of historical interest.

Tennessee Heritage Sampler Quilt

Dyes used in the Tennessee Heritage Quilt: **reds and pinks:** Turkey red, madder-black oak bark, cochineal, coreopsis; **yellows:** black oak bark, goldenrod, chrome, iron buff; **blues:** indigo, indigo-cochineal, indigo-iron tannate, Prussian blue, Prussian blue-logwood, Prussian blue-safflower, logwood; **greens:** indigo-goldenrod, indigo-chrome-yellow, Prussian blue-chrome yellow, Prussian blue-old fustic, logwood-old fustic; **purples:** madder, logwood, cochineal; **oranges:** madder-black oak bark, coreopsis, chrome, iron buff; **browns:** Bombay cutch, cutch-Osage orange, cutch-Osage orange-logwood, mineral khaki, manganese bronze, cutch-walnut, coreopsis-tannin; **grays and blacks:** iron tannate, iron tannate-mountain laurel, iron tannate-logwood, Prussian blue-mountain laurel, logwood-madder-black oak bark. Quilt designed and sewn by "Thursday Bee," Smoky Mountain Quilters of Tennessee. Dyed cotton fabric and photo by Jim Liles.

Cotton Handkerchiefs

Cotton dyes by Jim Liles from recipes in the text. Center is Turkey red, indigo blue, and chrome yellow. Photo by Jim Liles.

"Egyptian Purple" and Madder Reds

Cotton velveteen dyed "Egyptian Purple," bordered by madder red dyed cotton yarns. Wool felt flower is madder dyed; black silk scarf, Prussian blue-logwood. Dyes and photo by Jim Liles. Felted flower by Dale Liles.

Woolen Yarns

Goldenrod yellows, goldenrod-madder oranges, madder and cochineal reds, cochineal-indigo purples, goldenrod-indigo greens, indigo blues. Dyes and photos by Jim Liles.

Jim Liles and Red Yarns

Turkey red dyed linen shirt; Turkey red, madder, Brazilwood, and safflower red cotton yarns. Shirt sewn by Dale Liles. Dyed linen cloth and cotton yarns by Jim Liles. Photo by David Liles.

Turkey Red Shirt

Turkey red dyed cotton yarn and photo by Jim Liles. Cloth woven by Helen Gairns. Shirt sewn by Dale Liles.

Silk Scarves

Dyed scarves and photo by Jim Liles. Dyes used (top to bottom):

1	turmeric	Saxon blue-fustic	indigo	madder-black oak
2	black oak	indigo-turmeric	indigo	madder-cochineal
3	black oak	madder-black oak	indigo	madder-black oak
4	black oak	madder	indigo	madder
5	saffron	madder-black oak	Prussian blue	cochineal
6	saffron	madder	indigo	madder-cochineal
7	Queen Anne's lace	madder	Prussian blue	cochineal
8	gunstock	madder-cochineal	indigo	lac
9	cutch-indigo	madder	indigo	cochineal
10	cutch	madder	indigo	madder-cochineal
11	cutch	madder	Prussian blue	cochineal
12	cutch	Queen Anne's lace	Prussian blue	cochineal
13	black walnut	Queen Anne's lace	indigo	cochineal
14	cutch	saffron	Prussian blue	cochineal
15	cutch	saffron	indigo	cochineal
16	cutch	annotto	Prussian blue	cochineal-indigo
17	cutch		Prussian blue	Saxon blue
18	iron-tannate		indigo	Saxon blue
19	dye failure black		indigo	Prussian blue-logwood
20	cutch-logwood		indigo	Kelly green
21	logwood-old fustic		indigo	bottle green
22	Prussian blue-logwood			
23	Prussian blue-logwood			

Silk Scarf with Wool Felt Border

Scarf and wool dyed with cochineal-madder
combination by Jim and Dale Liles.
Finished scarf and felt by Dale Liles. Photo
by Jim and Dale Liles.

Mineral Dyes

Dyes used, counterclockwise from top: chrome oranges; iron buffs; iron tannate
grays and blacks; manganese bronze, khaki, and gunstock browns; chrome and
sea greens; chrome yellows. Center: Prussian blues. Cotton yarns dyed by Jim
Liles. Photo by Jim Liles.

Cotton and Silk-Fold Dyed

Scarves folded and/or shade dyed with indigo by Jim Liles. Photo by Jim Liles.

Felted Hat

Hand felted wool hat with silk band indigo dyed by Jim Liles. Felted hat by Dale Liles. Photo by Jim and Dale Liles.

Indigo Plant

Indigofera suffruticosa; specimen from our garden. Photo by Jim Liles.

Fur Pelt and Leather Shoes on Turkey Red Background

Pelt and leather indigo dyed by Jim Liles. Cotton cloth background dyed Turkey red by Jim Liles. Photo by Jim Liles.

1750s British Uniforms

Drum Sgt. Major Sheldon's woolen clothing is dyed cochineal scarlet (old process) and indigo overdyed with old fustic (popinger green); regimental cotton coat lining, Turkey red. Pvt. Liles' woolen clothing is madder red and popinger green; linen gaiters, cutch brown. Dyed items by Jim Liles. Photo by Jeff Wells.

Close Up of Uniform Mitre Cap

Woolen cloth dyed cochineal scarlet and popinger green by Jim Liles. Hat sewn by Ted Sheldon; embroidery by Kay Sheldon. Photo by Jim Liles.

Jim's Wool Felt Vest

Grey wool indigo dyed. The pattern is walnut brown and coreopsis rust; gunstock brown silk scarf. Wool and silk dyes by Jim and Dale Liles. Hand felted wool vest by Dale Liles. Photo by David Liles.

Constitution Competition Quilt

Dyes used: indigo and logwood blues; cutch, cutch-logwood-Osage orange, walnut and gunstock browns; madder purples; iron tannate and mountain laurel grays; black oak bark and iron buff yellows; Turkey and madder reds; madder-black oak bark orange; indigo-old fustic, chrome, and logwood-old fustic greens; cochineal pink. Quilt design, Linda Claussen; center quilting, Eva E. Kent. Sewn by "Thursday Bee," Smoky Mountain Quilters of Tennessee. Dyed cotton fabrics by Jim Liles. Photo by Blair White.

4 Blue Dyes

Superb blues may be obtained on cotton, linen, silk, and wool with logwood, Prussian blue, vatted indigo, and indigo sulfonate (Saxon blue). Logwood and Saxon blues are the easiest procedurally and are quite pretty newly made, but they possess only fair lightfastness and should be dyed in relatively deep intensities. Saxon blue is satisfactory only for wool and silk. Prussian blue, a mineral dye, produces lightfast shades that possess a brilliant sheen, particularly on fine linen, perle cotton, cotton sateen, and silk. Vatted indigo may be dyed from the palest blue to deep indigo purple-black, with many shades which are true indigo, i.e., between violet and blue. Indeed, it is possible to obtain ten or more shades of indigo, with all but the palest lightfast if properly done. The blues from the four dyes differ, and the experienced eye can easily distinguish a vatted indigo blue from Saxon blue, logwood, or Prussian blue.

Logwood Blues

Logwood was introduced as a dye to Europe after the discovery of America. It was one of the dye products taken back to Europe by the Spanish explorers (along with brazilwood, old fustic, and cochineal).

Botanically, logwood is *Haematoxylon campechianum,* a member of the pea family (Leguminoseae). It is a tree that grows to fifteen meters or so in size; the bark is thin and smooth, with thorns; its leaves resemble laurel; and the wood is very hard and compact, capable of taking a high polish (Napier, 269, and Brunello, 358). The specific gravity of the wood is greater than water, in which it sinks.

The best logwood came from the eastern shores of the Mexican Bay of Campechy, though it was also imported from Jamaica and other areas in subtropical and tropical America (Brunello, 197). Other names for logwood were *bois de Campeche* and *bois bleu* (Fr.), *palo-campechio* (Sp.), *Blauholz* (Ger.), and *legnotauro* (It.).

At first, the dye was not too well accepted since it faded badly with the older mordant methods (early sixteenth century). The government of Queen Elizabeth I of England banned its use in a statute titled "An Act for the Abolishing of Certeine Deceitful Stuffe Used in the Dyeing of Clothes," first enacted in the twenty-third year of Elizabeth's reign (1581). Approximately eighty to a hundred years passed before adequate experimentation indicated the tremendous qualities of the dye, and, ultimately, no natural dyewood saw more use. It was used in commercial dyeing longer than any other natural dyestuff, including indigo. Cheap woolens and silks were dyed black with it until about 1940 (Adrosko, 47). It is still used as a biological stain for certain animal tissues.

In 1765, the Spanish ship *El Nuevo Constante* sailed from Cadiz, Spain, bound for Vera Cruz, Mexico. Its cargo included items such as nails, plowshares, and wine, but the main item was mercury, which was used in extracting silver from relatively low-grade Mexican ores (the amalgamation process). The return cargo included Mexican-made ceramics, vanilla, leather, gold, silver, and copper, as well as two grades of cochineal (10,884 pounds), annotto (5,400 pounds), indigo (3,096 pounds), and *logwood (40,000 pounds)*. The logwood pieces were forty-eight to fifty inches in length and four to twelve inches in diameter (*El Neuvo Constante*, 33).

El Nuevo Constante sank in a storm off Cameron Parish, Louisiana, in 1766. The sunken boat was excavated in 1980 by the Louisiana Archeological Survey and Antiquities Commission. The cargo of logwood was in good condition in that the boat had been buried quite deeply in mud and water damage was thus kept to a minimum. Through the kind courtesy of Dr. Carl Kuttruff, Tennessee State Anthropologist, I have carefully examined one of the pieces of the retrieved logwood. The piece is nearly four feet long and four inches in diameter. Even after spending 215 years in ocean mud, it is extremely dense and hard. It appears to have been stripped of bark and possibly sapwood prior to being loaded on board ship.

The famous French dye chemist, Chevruel, studied the wood and named the dyeing principle *haematine* in 1812 (changed later to *hematoxylin*, Napier, 269). Haematoxylin exists in the wood as a colorless, crystalline substance. Upon exposure to air, it oxidizes to hematine. Chemically, haematoxylin has the empirical formula $C_{16}H_{14}O_6 \cdot 3H_2O$.

The wood was prepared for use by stripping off the bark and outer sapwood. Then the wood was chipped, moistened, and allowed to ferment. The fermented chips were boiled to extract the dye.

The dye was used with various mordants and at various degrees of acidity or alkalinity for shades of red, purple, blue, and black. The reds were dyed at low pH and were extremely fugitive; properly mordanted, the blacks were especially fast.

Logwood blues on cotton and linen are not as lightfast as indigo or Prussian blue, neither are they on wool, but they do fade fairly true, becoming lighter and slightly greener with time. However, they were much used in the past because logwood was so very much cheaper than indigo. At one point in 1831 logwood cost 6 cents a pound while indigo cost $2.25 per pound (Adrosko, 8).

Two recipes are given here for both cotton and linen and one for wool. These recipes work very well, yielding medium to deep shades of blue or greenish gray-blue. The first recipe is a modification of that found in Bronson (142); other old recipes are similar. All recipes are for one pound of material.

Greyish Green Amish Logwood Blue: Cotton or Linen

1. Scour the material well, rinse, and leave wet.

2. Add 3 to 4 gallons of water (hard water containing calcium is best) to a stainless, copper, brass, iron, or porcelain lined vessel. Heat the water to 160° to 180° F.

3. Add 1 1/4 ounces (3 level tablespoons) of impure blue vitriol (bluestone, copper sulfate), and stir until dissolved. Use only 2 level tablespoons of high grade copper sulfate.

4. Add the yarn or piece goods and mordant for 1 hour or more. Move the material several times during this period.

5. While the mordanting is under way, add 6 ounces of logwood chips or, preferably, 4 ounces of logwood sawdust to 4 to 6 gallons of water and heat to about 180° F. Keep the kettle at 180° F for about 45 to 60 minutes.

6. Allow the logwood liquor to settle and cool to about 120° to 140° F. Then decant the clear liquid and add the mordanted yarn (remove from the mordant and squeeze before adding the material to the logwood liquor).

7. Turn and move the material in the logwood liquor for about 30 to 40 minutes. Have the vessel off the fire during this time; the liquid will remain hot enough.

8. Remove the material, wring out, and hang it up.

9. Dissolve 2/5 ounce (3 level teaspoons) of washing soda or pearlash in the logwood liquor and reenter the goods, turning at intervals for another 20 minutes.

10. Remove, squeeze, and dry, without rinsing.

11. If the color is not dark enough, add 1/5 ounce or about 1 teaspoon of copper sulfate to the liquor, stir until dissolved, and reenter the material for another 10 minutes.

• A deep navy blue results with use of 6 to 8 ounces of logwood sawdust and 1 1/4 ounce of high-grade copper sulfate.

• Properly done, these blues fade far less than I had expected.

Logwood Navy Blue: Wool

This recipe is essentially from Bronson (179).

1. Add 4 to 6 gallons of water to a stainless, copper, iron, brass, or porcelain lined vessel and heat to about 180° F.

2. Remove the vessel from the heat and add 3/4 ounce (2 tablespoons) of copperas, 1/5 ounce (1 teaspoon) of alum, 1/10 ounce (1/2 teaspoon) of verdigris or copper sulfate, and 1/10 ounce (1 teaspoon) of cream of tartar. When all of these materials are dissolved, add the well-scoured and wetted-out wool, and heat to the simmer. Mordant at this temperature for 1 hour.

3. Remove the wool, squeeze, and hang the wool to air for 10 to 15 minutes.

4. Rinse well.

5. Add 4 to 6 gallons of water to the same or a similar vessel, and add 4 ounces of logwood chips or 3 ounces of logwood sawdust (either free or in a porous cloth bag), and simmer for 45 to 75 minutes. Remove the bag or decant the liquor after allowing time for settling.

6. Add 1/2 ounce of madder broken up fine (or in a little bag), or 1/2 gram of alizarin.

7. Add the wet, mordanted wool, bring up to the simmer, and keep at that temperature for 30 minutes.

8. Remove the wool, squeeze, and air.

9. Dissolve 2/5 ounce (2 teaspoons) of copper sulfate and 1/5 ounce or 1 teaspoon of washing soda or pearlash to the dyebath, and stir well.

10. Reenter the wool and heat back up to the simmer. The wool should remain in the dyebath this time for 20 to 30 minutes.

11. Remove, squeeze, air, and then rinse.

• This recipe does seem to have a number of mordants; the effect is to produce a far more lightfast product than otherwise. The function of the madder (red or violet) is to offset the slight greenish tinge the logwood blue would otherwise assume, in time.

Logwood-Indigo Blue: Cotton or Linen

This recipe is essentially from Napier (369).

1. Scour well and dye the wetted-out goods a light indigo blue (any vat indigo method is satisfactory).

2. Mordant with tannin at the rate of 3 to 4 ounces of sumac leaves or 1/2 to 1 ounce of tannic acid for several hours. Dissolve the tannic acid in 4 to 6 gallons of water at 160° to 180° F, then add the cotton. Allow the material to remain in the dyebath as it cools.

3. Add 1 1/2 liquid ounces of acetate of alumina (red liquor) and 1 1/2 ounces acetate of iron (iron liquor) to 4 to 6 gallons room-temperature water. Then add the wrung-out cotton from the tannin and work well for 15 to 30 minutes.

4. Rinse well from the iron-alum mordant, and work for 20 to 30 minutes in a dyebath (about 140° F) made by simmering 6 1/2 ounces of logwood chips or 4 ounces of logwood sawdust for 45 to 60 minutes. Work the material in the dyebath for 20 minutes; remove the material briefly (lift and raise), add 3/4 to 1 liquid ounce acetate of alumina; reenter the goods; work 10 minutes longer; wash and dry.

• See Appendix B for production of "red liquor" and "iron liquor." One and a half ounces of Basic Alum Mordant No. 2 and 1/2 teaspoon copperas may be substituted.

Indigo Substitute: Cotton

Hummel (325) discusses, briefly, this material which he described as a purplish blue liquid produced by boiling together logwood extract and chromium acetate. Apparently, cotton was dyed directly by a hot solution of the mixture diluted with water. I have not tried the method.

Prussian Blues (Napoleon's, Berlin, Turnbull's, Paris, Raymond's, and Chinese Blue)

Fermentation vat indigo was the only really permanent blue until Macquer (French) first produced this mineral blue dye on silk in 1749. Following his discovery, the dye went under several names. With proper dyeing it produces stunning blues, greenish blues, and reddish blues on cotton, linen, silk, and wool. It is particularly beautiful on silk, for which it was much used. It is also quite striking on high-quality cottons, perle cottons, and fine linen. Fermentation vat indigo blue rarely gave beautiful blues on silk, and I would imagine that Prussian blue eventually took all Paris by storm. An excellent account of the discovery of the dye is given in Brunello (231) and in Pellew (1912).

Prussian blue pigment was produced first by Diesbach, a Berlin manufacturer of paints, quite by accident, in 1704. It was the result of impurities in his materials. However, he described the color, and in 1724 Woodward published the method for producing the deep blue. In 1749, Pierre J. Macquer, Assistant to the Chair of Chemistry at the Jardin des Plantes, discovered the process of dyeing silk directly using a combination of iron (ferrous) salts and potassium ferrocyanide.

Macquer published his results ("L'Art du Teinturier en Soie") in 1759 (Brunello, 231), but this may not have gone beyond France because Pellew (1912b) gave a different, and later, account of the discovery of Prussian blue approximately as follows:

> Prussian blue was discovered accidentally by two young German chemists in 1788. They sought refuge at a country inn during a rainstorm. Behind the inn stood a blacksmith's shop, which they visited after the rain had stopped. They found a deep blue puddle in a depression on the paving stones. They soon discovered that the blue color developed whenever a trickle of water from a pile of rusty horseshoes in one corner of the yard came in contact with water draining from a pile of ashes from the forge. The ashes contained not only burned charcoal, but also scraps and shavings from the hooves of horses. They later determined that the "Prussian Blue" resulted from the combination of iron salts with a yellow salt, potassium ferrocyanide (yellow prussiate of potash), formed by the action of fire on the potassium carbonate in the wood ashes with the nitrogen of the hoof scrapings and the iron from the scales of the horseshoes. (93)

Prussian blue was very extensively used on cotton, wool, and silk during all of the nineteenth century; it was especially employed for army uniforms (Matthews, 523). By the beginning of the twentieth century it was largely replaced by the then-popular synthetic alizarine blues. Alizarine blue was by that time very good, and it was brilliant, cheap, and easier to apply than Prussian blue, or indigo for that matter. Also, Prussian blue is decomposed by hot alkaline solutions, and turn-of-the-century laundry soaps were often just that. By contrast, most of our modern detergents are nearly neutral and therefore not damaging to Prussian blue.

Prussian blue is very fast to light and exposure, and fairly fast to reasonably careful washing. Should it ever fade, several days in the dark will restore the original color. In very heavy shades it may not be entirely rubfast, particularly on wool. It is also fast to dilute acids, but is decomposed by strong acids. As previously mentioned, it is also decomposed by strong alkalis, leaving iron buff. This latter reaction was used for discharge work in printing cloth at least as late as 1916 (Matthews, 523). Prussian blue is still used as a pigment and as a test for the presence of iron (for the traditional dyer in samples of alum).

Not only was Prussian blue used as a self color on all the natural fibers, but also as a "bottom color" in dyeing feathers and silk full shades of black. In addition, Prussian blue overdyed with chrome yellow produced the green dye of our "greenback dollars" from about 1850 to 1900 (Pellew, 1912b p. 94). Finally, Prussian blue overdyed with safflower produces a stunning lilac, a dye that was listed in dye manuals as late as 1885.

Caution: Production of Prussian blue results in release of tiny amounts of toxic hydrogen cyanide gas. Therefore, it is suggested that these dyes be made outside. The Prussian blue itself is nontoxic.

All recipes are for 1 pound of material.

Sky Blue Recipe No. 1: Cotton, Linen or Silk

The two most common shades of Prussian blue were "sky blue," and "royal blue." Royal blue requires a deeper base of iron buff and somewhat greater amount of yellow prussiate.

1. Dye the material a light iron buff (see recipe for iron buff, Orange chapter).

2. Dissolve 5 to 10 teaspoons of potassium ferrocyanide in 5 gallons of warm to room-temperature water. This can be done in a plastic paint bucket or waste basket.

3. Add the well-wetted out buff-dyed yarn or piece goods and work in the solution for 3 to 5 minutes.

4. Briefly remove the goods and add about 1/2 ounce of concentrated acid (any kind) or 6 to 8 ounces of good vinegar. Stir, reenter the goods and work vigorously for the next 2 to 3 minutes. If the solution is acidic enough, the color will develop instantly. If the blue color does not develop almost instantly upon entering the material into the acidified prussiate, add more acid or vinegar.

5. After working the material 2 to 3 minutes, let it remain in the solution for 15 minutes.

6. Next, work the material in a cold solution of 1 ounce of alum per pound of cotton in 4 to 5 gallons of water (use a boiling solution of alum if the material is dyed unevenly).

7. Squeeze, hang up, and dry.

8. One or two days after drying, wash gently in mild detergent, hang up, and dry (optional).

• Royal blue requires, initially, dyeing of a deeper iron buff. The final color is more dependent upon the depth of iron buff than on the amount of prussiate. However, use half again more prussiate for royal blue.

• For tie-dye work where both Prussian blue and iron buff are wanted, dye the full blue, then tie, and discharge away the blue in the untied portions, or dye iron buff and then tie and dye the untied portions the blue. To discharge the blue back to iron buff, place the item in a hot (140° to 190° F) solution of soda ash (washing soda). The solution may be made by adding 1 heaping teaspoon of soda ash per gallon of hot water.

• If you wish to discharge the iron buff as well, great care must be exercised because the process requires strong acid (about a "1 normal" solution, pH less than 1). To prepare the solution, carefully and slowly pour 1 part of concentrated hydrochloric acid (muriatic acid) into 11 parts of water in a glass vessel or mix (11 to 12 ounces of acid into nearly a gallon of water). *Never pour water into acid.* The preparation of the acid solution should be done outside or in a fume hood. Next, work the wet material with glass or plastic rods for a few minutes in the solution. Then pick up the material with rods and rinse a minimum of 3 times. One rinse with alkaline washing soda solution or diluted ammonia will remove all traces of acid, which is mandatory with cellulosic materials. If strong mineral acids are permitted to dry on cellulosic materials, the fibers will be destroyed, or at least greatly weakened.

• Use of nitric acid, instead of sulfuric, hydrochloric, or acetic often imparts a teal shade to the material. In concentrated form, nitric acid releases quite noxious fumes — much less if the acid has been diluted.

Sky Blue Recipe No. 2: Cotton, Linen, or Silk

This recipe is essentially from Napier, and is typical of nineteenth-century recipes.

1. Scour the material well, and bleach if necessary.

2. To 4 to 6 gallons of room-temperature water add 1 liquid ounce of nitrate of iron and stir (see Appendix B). Late in the nineteenth century basic ferric sulfate (Monsells salts) was substituted for "nitrate of iron." Here the iron is already in the necessary ferric state and may be used by simply dissolving it in water. This avoids preparing the "nitrate of iron" from iron and nitric acid. Ferric chloride may also be used (Matthews, 519). Use 2 to 3 teaspoons per pound of material as a starting point.

3. Add the material and work well for 15 to 20 minutes.

4. Wring out and rinse slightly in a tub of clean water.

5. In another tub of room-temperature water (4 to 6 gallons) dissolve 1/2 ounce (5 to 6 teaspoons) of yellow prussiate of potash. When dissolved, add about 1/2 liquid ounce of oil of vitriol (concentrated sulfuric acid) or 8 ounces of vinegar. Pour the acid into the liquid slowly, while stirring.

6. Add the goods and work in the solution for 15 minutes.

7. Wring out and wash through cold water in which is dissolved about 3/4 teaspoon of alum.

8. Wring out and dry without rinsing.

• To dye lighter or darker shades of sky blue, use more or less of the iron and yellow prussiate; or should the shade be too light, repeat the operations through the same tubs, but add 1 teaspoon more of the yellow prussiate.

Napoleon's and Royal Blue: Cotton, Linen, or Silk

This recipe is of unknown origin, though essentially the same as that found in many nineteenth-century manuals.

1. Add 6 fluid ounces of nitrate of iron (see Appendix B) or several teaspoons of ferric sulfate or ferric chloride to 4 to 6 gallons of room-temperature water.

2. Then dissolve 3/5 ounce (3 teaspoons) of stannous chloride in a solution made by pouring, carefully and slowly, 1 1/2 liquid ounces of hydrochloric acid into 3 ounces of water. Add this solution to the iron bath and stir.

3. Enter the scoured and wetted-out yarn and work the material for 30 minutes.

4. In another tub dissolve 1 1/2 ounces (4 tablespoons) of yellow prussiate, and when dissolved, add 1/2 liquid ounce concentrated sulfuric acid (add the acid slowly and carefully, with stirring).

5. Remove the goods from the iron-tin bath, squeeze, and place in the prussiate bath. Work for 15 minutes.

6. Remove, squeeze, and rinse in a solution made by dissolving 1/2 ounce (3 teaspoons) of alum in 4 gallons water.

7. Remove, squeeze, and dry.

• Repeat the procedure, using the same tubs if the color is uneven or not dark enough.

• Use twice as much iron and 1/4 more tin and prussiate for royal blue.

Prussian Blue Recipe No. 1: Wool

Wool may be dyed Prussian blue with potassium ferrocyanide (yellow prussiate) or potassium ferricyanide (red prussiate). Recipes using red prussiate are in most cases superior, but yellow prussiate is also satisfactory. Light to medium shades on wool are recommended; dark shades are apt to crock slightly and tend to make the wool harsh to the touch.

1. Prepare a dyebath (in any type of vessel capable of being heated) by dissolving 1 1/2 to 1 3/4 ounces (4 tablespoons) of red prussiate in 4 to 6 gallons of room-temperature water.

2. Next, add carefully, and with stirring 1 1/2 liquid ounces of concentrated sulfuric acid.

3. Add the well-scoured and wetted-out wool, and gradually raise the temperature of the bath to the simmer over the course of an hour. Keep at the simmer for 30 minutes longer (the wool will turn green first, and then blue).

4. If 1 to 1 1/2 teaspoons of tin (stannous chloride) is added at the beginning (after adding the acid) or at the beginning of the last half hour, the color will be brighter and somewhat reddish purple.

5. Remove, cool, rinse or wash well with detergent, and dry.

• Acids other than sulfuric may be used, which will affect the shade slightly. Nitric acid makes the shade somewhat greener. Remember that concentrated nitric acid is fuming and nasty, and should be poured outside or in a fume hood.

Prussian Blue Recipe No. 2: Wool

1. Dissolve approximately 3 to 3 1/2 ounces (8 tablespoons) yellow prussiate in 4 to 6 gallons of water (in a suitable vessel that can be heated).

2. Add 2 level teaspoons of alum and 2 level teaspoons of cream of tartar. Stir until all materials are dissolved.

3. Add the well-wetted-out and scoured wool and heat up slowly to the simmer, and keep at that temperature for 15 minutes.

4. If dark enough, remove the wool and proceed to next step. If not dark enough, add additional chemicals in the same proportions but smaller amounts to the same dyebath, reenter the wool and simmer for another 15 minutes.

5. Remove, cool, rinse, wash well with detergent, and hang up to dry.

Indigo Blues

With the possible exception of iron buff and the tannins, indigo has probably seen longer continuous use than any other dye (perhaps 5000 to 5500 years). Possibly the oldest dated specimen is a linen fragment from Thebes, ca. 3500 B.C. (Matthews, 8), and indigo is still being used for faded blue jeans, the finest Japanese kimonos, and specialty items for fiber artists. Since indigo penetrates slowly, it is an ideal dye for fold or stitch dye work (*plangi, shibori, adire'*).

Without indigo-bearing plants, the world prior to about 1650 (and until much later in most places) would have been much drearier, because nature is otherwise sadly lacking in fast natural blue dyestuffs.

Vatted indigo, except in the palest shades, is quite lightfast, resistant to acids and alkalis, and is washfast, sweatfast, and – properly dyed and scoured – rubfast. One other notable property of indigo is the fact that it often becomes even clearer and more beautiful with successive washings. This may be more true of the natural product than of the synthetic.

Over fifty species of the families Leguminoseae, Cruciferae, and Apocynaceae possess sufficient dye to have seen use. The principal species are listed in Table 1 (from Gerber, 1977, 8).

Because indigo-bearing plants grew indigenously over wide geographical areas, the dye was used by many cultural groups at very early dates – in a number of places by 2000 B.C. Their success is all the more impressive in that indigo is not a simple dye, and its chemistry was not elucidated until quite late (1750–1800). Therefore, before 1750 indigo dyeing was all art, technique, and experience. For many primitive cultures it appeared to possess magical qualities with religious overtones, perhaps because indigo and royal purple are the only natural dyes which develop their colors after removal from the dyebath (by oxidation in the air).

Table 1. Origins of Indigo

SPECIES	COUNTRY OF ORIGIN
Indigofera tinctorium	India
Indigofera suffruticosa	Mexico and South America
Isatis tinctoria (woad)	Europe, Egypt
Lonchocarpus cyanescens	West Africa
Marsdenia sp. (milkweed)	Sumatra
Nerium tinctorium (oleander)	India, Far East
Polygonum tinctorium (buckwheat)	Central Asia, China, Japan

Indigo: General Features

Indigo dye (indigotin) is derived from the glucoside indican, $C_{14}H_{17}NO_6 \bullet H_2O$, which is a soluble, colorless substance present in the indigo-bearing plant. Indican is composed of glucose (dextrose, $C_6H_{12}O_6$) and indoxyl, C_8H_7NO.

When indican-bearing plants are crushed and placed in water, the glucoside (indican) is released, and within a few hours bacterial enzymes hydrolyze (remove) and consume the glucose, leaving the indoxyl radicals. If textile fibers are placed in the solution, left for a time, and then removed, pairs of indoxyl radicals plus oxygen combine forming indigo blue (indigotin, $C_{16}H_{10}N_2O_2$). Indeed, such treatment dyes the fiber a very pale indigo blue. Or if air is beaten into the solution, indigo blue results by the same reaction, and the insoluble indigo blue eventually settles out, concentrating the dye. The chemical reaction is:

$$2 \text{ INDOXYL} + O_2 \rightarrow \text{INDIGOTIN} + H_2O$$

Indigo blue (the oxidized form) is insoluble in water, weak acids and alkalis, and will not dye anything permanently. First, indigo blue must be reduced chemically to indigo white or leucoindigo, in which form it is soluble in an alkaline solution. This chemical reaction occurs as follows:

INDIGOTIN (insoluble) + reducing agent and alkaline solution → INDIGO WHITE (soluble)

Fiber immersed in such a solution (the solution is amber to yellowish green) will be penetrated by the soluble indigo white. In this state the soluble indigo white makes loose chemical combination with the fiber molecules. Now, if the fiber is taken from the vat into the air, the indigo white oxidizes back to the insoluble blue form — the magic — and remains in relative permanence. The indigo blue, at this point, is both mechanically held within and on the surface of the fibers, and weakly combined chemically by hydrogen bonding and Van der Walls forces.

The first indigo dyeing probably involved simple use of the crushed, fresh plant material solutions. I have experimented twice with this method with good results. Good directions are given by Buchanan (1987a, p. 118) and Gerber (1977, p. 27). Bancroft discussed his experiments using the direct plant method in his second volume, published in 1814. In addition, I strongly suspect the following accidental process of discovery to have occurred in many places and very early. Soap, as we know it, was not widely known until about the first century A.D. Thus, other than suds-bearing plants, the only cleaning materials available were alkaline liquids such as ammonia, which develops by bacterial action on urea in old urine, lime water, and the potassium carbonates and hydroxides gotten from running water through hardwood ashes (wood ash lye). These materials, particularly urine, were collected and stored by literally all primitive cultures. If, by chance, some indican-bearing plant material found its way into a urine vat, the bacteria growing therein would render the vat in a reducing condition (use up the oxygen and release hydrogen), and the ammonia, being alkaline, would dissolve the resulting indigo white. Now, if some fiber fell into the vessel, and if it were retrieved later, the yellowish green material would turn blue before the eyes of the retriever, and the material would be dyed a permanent very pale blue.

All of the earliest recorded indigo dyeing was done using stale urine, lime water, or wood ash lye for the alkali, and microorganisms for the reducing agent. Indeed, lime water, urine, or wood ash lye and microorganism vats were regularly used with natural indigo in such places as Africa and the southern Appalachian mountains of the U.S. well into the first part of this century, and the traditional dyers of Japan and elsewhere still use this method.

Indigo may be dyed from the palest shades to almost a purple-black. Bemis (108), listed thirteen shades of indigo blue, lightest to darkest as follows: milk, pearl, pale, flat, middling, sky, queen's, turkish, watchet, garter, mazareen, deep, and navy.

Of course, these shades were named using natural indigo, which often gives slightly more subdued or earthy tones than synthetic indigo, due primarily to plant impurities. The natural product also contains variable amounts of indirubin (red indigo), indigo gluten, indigo brown, and mineral matter. Practically all of these materials are either destroyed in the dyebath or are eliminated in the wash waters, thus having little effect on the color (Matthews, 410). The greater the percentage of impurities, such as red indigo, the harder the natural product will be, making it more difficult to grind to powder. Also, red indigo reduces more slowly than normal indigo; thus natural indigo with a high indirubin content is harder to reduce (Matthews, 410–11).

Synthetic indigo is chemically identical to natural indigo (indigotin) and was first synthesized by Adolph von Bayer in 1880. By 1897 improved methods of production permitted commercial competition with the natural product, and by 1920 synthetic indigo had almost completely replaced the natural product (Matthews, 432). It could be made more cheaply, about half the price of the natural product, and was more uniform in concentration than natural indigo. The best natural indigo contained up to perhaps 70 percent indigotin, but some of the poorer grades were, and are, as low as 20 percent.

In a way, it is a miracle that natural indigo is still available at all, but I am certainly glad that it is because its use is sometimes desirable for sentimental or even for practical reasons. For instance, fermentation vats sometimes do not work with synthetic indigo: perhaps traces of some chemical used in the making of the synthetic product poison the microorganisms. Also, sometimes the slightly more subtle tones of the natural product are desired. And I would not want to use synthetic indigo to produce replacement material for, say, a worn old quilt made before 1900.

In 1964 Dale Liles purchased 2 pounds of natural indigo from Skilbeck Brothers of London. This company was in business from 1650 to 1970, a period of 320 years. They were purveyors of the finest natural dyestuffs in the world, and their natural indigo, probably from Bengal or Java, was without equal. It was refined to about 60 percent, a soft crumbly substance called "indigo vat grains." To prepare it in paste form requires virtually no grinding, and it reduces beautifully, regardless of the reducing agent. In short, it was, and is, easier to work with than any other form of indigo, natural or synthetic, except for a professionally prepared synthetic paste. A few of us are fortunate enough to still have some of this exquisite material. We use it with discretion, and only for special projects.

Fortunately, two types of prepared natural indigo are still available, *I. tinctorium* from the East, and *Polygonum* sp. from Japan. The better grades of *I. tinctorium* are dark blue, somewhat earthy-looking lumps, not too hard; when scratched with the fingernail they show a coppery streak. The poorer grades, of low indigotin content, are usually hard stony lumps, blue, but appearing to contain mineral material, requiring tedious grinding – or soaking and grinding – before use. The Japanese product comes in the form of a soft, moist patty which contains considerable plant material. Good natural indigo is elegant and feeds one's atavistic tendencies. It will remain on the market only if you and I purchase and use some from time to time.

Dyeing well with indigo is not the impossible bogeyman often depicted. However, it does require proper instruction, experience, commitment, and comprehension of what is going on in the dye vat. One of the major problems is that the vast majority of recipes in current dye manuals are incorrect as to proportions, usually calling for excessive lye or other alkali and too much indigo, especially for wool dyeing. Excess alkali causes deterioration of protein fibers, makes them harsh to the touch, and enhances felting. In addition, excess alkali makes the indigo white so soluble that its affinity for the fiber is decreased, making it more difficult to build deeper shades by redipping. Deep shades of indigo are always produced best by successive redipping in a weak to moderate vat, but if the vat is too alkaline often as much previously deposited indigo is re-reduced and stripped from the fiber as is added. A too-alkaline vat also often produces a poor color, inclining towards the blue-gray. Fortunately, this can usually be corrected by an acidic (vinegar) afterbath.

If one finds one has too much indigo in the dyebath, it may be the result of copying old recipes originally designed for natural indigo, which always has a lower indigotin content than does the much more commonly used synthetic. Synthetic indigoes are not 100 percent indigotin either, but they do have a higher dye content than any of the natural forms. Furthermore, natural indigoes vary as to percent indigotin content. Many contemporary dyers wish to do a speedy job – that is, have enough indigo in the vat to get a deep shade with one dip into the vat. The desired depth of color may be achieved in this way, but the results are often mediocre at best.

Since natural and synthetic indigoes vary as to indigotin content, and since some reduce more completely than others, it is not possible to give an absolute recipe for a given shade of indigo for number of pounds of fiber. Instead, the first dip should produce, usually, a lighter shade than that desired, which may be achieved by judicious redipping.

Another general problem encountered with vatted indigo dyeing is inadequate fiber preparation. Indigo penetrates slowly, and every method must be used to encourage it to do so. It is of paramount importance to have the material entirely grease free. It is futile to try to dye "spun in the grease" woolen yarns without thorough scouring. The indigo will not penetrate the fiber deeply, will never be rubfast, and, being a surface effect only, will likely fade. Hard-surfaced cotton or linen piece goods or yarns are also difficult to penetrate with indigo unless all traces of waxes, pectic substances, and oil have been removed by thorough scouring.

Excepting the fresh plant method, indigo may be dyed two different ways. The oldest is the vat process, which involves changing insoluble, oxidized indigo blue into reduced indigo white by reducing the blue form in an alkaline solution. This method is suitable for the dyeing of all natural fibers. This method is the more complicated and difficult to learn of the two but is the more satisfactory in that the product is more lightfast and washfast. The second method was discovered around 1740 or a little later, when concentrated sulfuric acid became available. This method involves producing a compound by reacting indigo with concentrated sulfuric acid. Such reaction produces indigo mono- and disulfonic acids. This compound is soluble in water and dyes directly and easily on wool and silk. A neutralization process is necessary for using it on cotton; this process was not very satisfactory or much used. The compound is an acid dye and went under the names of "Saxony blue," "chemic," "indigo carmine," "indigo sulfate," and "indigo extract." It gives good teal blue colors on wool and silk, though a different blue from that of vatted indigo, and very striking greens when overdyed with old fustic. The colors are not as permanent as with vatted indigo.

Because the process for vatted indigo is complex and often discouraging for the novice, and because each recipe has its unique difficulties, I have included a section on troubleshooting for each step where trouble is likely to occur.

Recipes for Vatted Indigo

Recipe No. 1: The Lye-Hydrosulfite Vat

This is the easiest recipe, though the latest method, historically. It is designed for approximately 1 pound of cotton, silk, or linen medium blue or 2 pounds light blue. The same amount of indigo will dye twice as much wool to the same intensity.

Preparation of the concentrate, or stock solution:

(Indigo is reduced and dissolved more easily and quickly in concentrated form, and the concentrate may be kept on hand until needed). In a wide-mouth quart mason jar nearly filled with hot tap water (110° to 140° F), dissolve 1 to 1 1/2 level teaspoons of lye (sodium or potassium hydroxide). Add 2 teaspoons of fine synthetic indigo powder or 2 to 4 teaspoons of well-shaken indigo paste (see below) and stir well for about 2 minutes. Then add 2 teaspoons of sodium hydrosulfite or 1 teaspoon of thiourea dioxide and stir for about 1 minute. If reduction of the indigo starts properly, the color of the surface of the liquid should change to a purplish violet with a coppery sheen. Now place the lid on the jar and set aside if the room is warm (summertime) or put the jar in a pail of hot tap water. If all proceeds normally, the liquid should change color from blue to muddy bluegreen to clear amber (yellow or yellowish brown) in 15 to 60 minutes. The indigo blue is now reduced to indigo white and is dissolved in the alkali. The surface of the stock may remain blue because the surface indigo is in contact with air. The stock may be used at this point or kept for an extended period. If not used, the indigo will eventually reoxidize, even with the lid on the jar, but this will take some weeks. At that point, the liquid should be heated again to 110° to 140° F, stirred well, and more reducing agent added.

Trouble-shooting the stock solution:

• If the stock does not reduce in 15 to 60 minutes, it may do so overnight. The solution does not have to be absolutely clear, but if it is not, this is usually because the indigo in the stock solution was not ground fine enough. Indigo of only very small particle size will reduce quickly, and above a certain particle size it will not reduce at all. Another possibility is that the hydrosulfite or thiourea dioxide are no longer effective. Maximum shelf life for these is 1 to 2 years.

• Sometimes indigoes of slightly larger particle size will reduce better with sodium hydrosulfite than with thiourea dioxide. To test this, heat the solution again (place the stock bottle) in a pan of water on the stove) to 120° to 140° F and add (with stirring) 3 teaspoons of sodium hydrosulfite. This compound may be obtained from dyestuffs suppliers or purchased at the store (it is the major component of Rit or other dye remover). The only problem with hydrosulfite is that it has a shelf life of only about one year, whereas thiourea dioxide, if kept dry, will last much longer. Thiourea dioxide is more expensive than hydrosulfite, but only about one-third as much is required (by volume).

• If the stock still is not clear and amber, but is bluish green or brownish green, this is an indication that some, but not all, of the indigo is reduced. It may be used this way, but there will be more unreduced indigo to wash out of your goods following dyeing. Thus, more of your indigo and time will have been wasted.

• Be careful not to heat the stock solution to a temperature higher than 140° F. Indigo, in reduced form, is destroyed by excessive heat. This is not true of oxidized indigo.

• Presence of a greenish yellow stock and white precipitate at the bottom of the stock jar is an indication that there is insufficient alkali to dissolve the indigo white. Addition of a teaspoonful of lye, with stirring, should correct the problem. Add only as much alkali as is necessary to dissolve the indigo white.

• Presence of a dark precipitate at the bottom of the stock bottle is an indication of (1) plant and mineral material (natural indigo), or (2) indigo which is not ground fine enough to reduce, or (3) non-indigo material of a chemical nature (even synthetic indigo is not 100 percent indigotin). In any event, when adding stock to the vat, do not pour in any precipitate at the bottom of the stock bottle. Instead, pour the clear into the vat; to the residue in the bottom, add hot water, lye, and reducing agent, as before, but add no new indigo. Stir well and see if some of the residue reduces. If little does, the residue should be thrown away.

Preparation of indigo paste:

Use fine synthetic powder. Quality of synthetic indigoes differs, both from the standpoint of the percentage of indigotin content and that of how finely ground they are (how small the individual particle size). Of course, there will be some variation in particle size in all, but the smaller the better. Ideally, the particle size should be so small that the material flies! Indigo this fine may be made into a paste easily without grinding, and after trying several methods I have found the following method (Mairet, 52) to work the best: Place boiling water (3 ounces for each 2/3 ounce or 2 level tablespoons of indigo for a 20 percent paste) in a pint canning bottle and place the canning bottle in a pan of boiling water which has just been removed from the stove. Spoon the dry indigo powder quickly into the jar, stir briefly, and replace the lid. The hot water and steam will wet the indigo powder very well. When cool, the paste may be stirred again, and it should always be shaken well before using. Commercial indigo pastes were and are usually 20 percent.

Natural lump indigoes and synthetic indigoes of large particle size should be ground before being made into a paste. (The exception is natural lump indigo, which may be placed in a very, very fine mesh bag or a bag made of extremely fine fabric and placed directly in a fermentation vat.) Each day or twice a day the bag may be rubbed, releasing into the vat the very fine indigo which will reduce. By the time the vat is in order, most of the indigo will have been released.

Reading descriptions of indigo-grinding in the old days is most illuminating. It was usually performed for hours in an iron ball mill, using conical shaped balls or common cannon balls (Bemis, 107). A home method involved suspending a 6-gallon iron bucket (with legs) by a rope, adding the indigo and some water and two 18-pound cannon balls. The grinder grasped one of the legs and shook the bucket so as to roll the balls in circular fashion. After some time the grinding was stopped, time was allowed for the heavier, yet unground material to settle, and the clear suspended blue was poured off. Then more water was added and the grinding repeated. The process was continued until all of the indigo was ground. The remaining residue was discarded. The modern equivalent, using smaller amounts, may be done in a reasonably large size (3- to 5-inch diameter) porcelain or iron mortar and pestle. A mortar and pestle may be purchased from chemical or biological supply houses, through your druggist, or sometimes from an antique dealer. The mortar and pestle is also very handy for grinding cutch, alum, etc. Approximately one ounce of indigo may be ground at a time in this size mortar and pestle, using the method of grinding, allowing time to settle, pouring off the clear blue water, and repeating the process. A well-ground 20 percent paste will feel as slick as oil between the fingers. (When grinding indigo, note how much indigo you end up with in what volume of water, e.g., 1 1/2 ounces in one quart.)

If no mortar and pestle is available, put the indigo in the finest weave cloth bag available and place this in a small volume of water. Then rub the bag with the fingers at intervals. Only the very fine indigo should end up in the water. This process should be carried out over several days if all of the fine indigo is to be gotten out of the bag.

Preparation of the indigo vat:

1. Unless you are going to dye only very small quantities, use the largest vessel available. I prefer at least a 5-gallon vat for 1-pound quantities, and 10 gallons is better. The larger vat permits the material to float freely and is easier to keep in order over a long period. Sometimes 2 yards of piece goods may be handled in a 10-gallon vat; a 1-yard piece requires at least a 5-gallon vat, and a 10-gallon or larger vat is better.

2. For one pound of yarn use a 5-gallon or larger copper, stainless steel, porcelain-lined, or plastic container (plastic garbage pails work just fine) and fill it nearly full with hot tap or deionized water (110° to 140° F).

3. Add 1/8 level teaspoon (for 5 gallons) or 1/4 teaspoon (for 10 gallons) of lye to the water and stir until dissolved. This renders the vat slightly alkaline so that the reduced indigo from the stock does not reoxidize when added. Lye is messy stuff and quite hygroscopic. I find it preferable to make up a stock solution of it as well; 1 teaspoon lye per ounce of water. This should be kept in a good, shatterproof plastic bottle with a tight lid.

4. Add approximately 1/2 level teaspoon of detergent or Turkey red oil per 5 gallons of fluid. This helps the indigo to penetrate and breaks up oxidized indigo on the surface of the vat.

5. Add 1 level teaspoon of thiourea dioxide (5- to 10-gallon vat) or 2 teaspoons of sodium hydrosulfite and stir very gently until dissolved. Put a lid (if available) on the vessel to exclude air and allow at least 15 minutes for the vat to be reduced.

6. Now, carefully add all of the contents of the stock bottle if you are dyeing cotton, linen, or silk, and half that much for wool. This is best accomplished by lowering the jar into the vat and then pouring out the contents, so that less air is introduced. Stir gently and allow several minutes to an hour, until the vat is yellowish green in color (test by taking up a white styrofoam cupful and checking the color). Never rush the vat.

Dyeing procedure:

1. Introduce carefully, and without a lot of splashing, the well-scoured and wetted-out material, or dry cotton, linen, or silk. Wool should always be wetted out. For very level dyeing, work it through the vat gently, with slight agitation at intervals. To insure complete

penetration, I prefer to allow at least 10 minutes for this first dip, and longer, say 30 minutes, for hard cotton or linen yarns. This also insures complete displacement of water or air from the material. Such a long dip is not required for very sheer silks, thin cotton batiste, or woolen yarns. The long time interval does not develop too deep a color on cotton and linen, unless the beginning vat concentration is excessively strong.

2. Now put on rubber or plastic gloves and squeeze the material out underneath the surface as much as possible and lift out over the grass or over a basin. Open the material up and give it the air! The color change from yellow-green to blue is fascinating to watch and always gives pleasure. If rubber or plastic disposable gloves are not used, the hands, and especially the fingernails, will become quite blue before the dyeing session is over. Alkali is hard on the hands, too. Prior to the invention of gloves, the dyer had a pot of clear or soapy water sitting next to the indigo vat, and the hands were rinsed as soon as possible after handling the material.

3. Be careful as to where the article is permitted to air. Never place the partially oxidized goods, particularly piece goods, over a cotton or other absorbable type of clothesline – or a line may appear on the article. Nonabsorbable plastic clotheslines are usually all right, as are hosed-off bushes or dry grass. Turn or move the article occasionally, and continue to squeeze out.

4. Amount of time essential for airing (oxidation) varies with the hardness of the material and temperature. Generally speaking, cotton and the harder linen require more time than the more open and porous wool. In some primitive, and not so primitive, cultures cotton goods may even be made into garments and sold in the unwashed or partially washed condition, the idea being that indigo may continue to oxidize for up to a year after dyeing. (Indeed, a traditional paper-maker friend unknowingly purchased such a blouse in Japan and ended up with blue armpits.) This long oxidation, in my view, is completely impractical. Indigo vat fluid may be allowed to dry on cellulosic fibers but should never be permitted to do so on protein fibers, since the alkali becomes more concentrated as drying proceeds. The one exception to this is if the vat was rendered alkaline with ammonia, which evaporates along with the water. In my opinion, overnight is plenty long in all cases, if the material is not to be redipped, and 10 minutes to 30 minutes is ample if the material is to be redipped. Twenty minutes is a good average, and if the item is wool or silk and it starts to become noticeably drier before

that time elapses, it should be placed in room temperature to cold water.

5. Finishing off: Wool and silk will reveal the true final color while still slightly damp, but cotton and linen appear 2 or 3 shades darker when wet, and it will be necessary to allow the material to dry completely before the true shade becomes apparent. If the shade is correct when dry, and the color even and level, the material may be left overnight or washed in an hour or so (remember that wool and silk should be put in water before they dry out).

6. Washing: All indigo vats contain some oxidized blue indigo which will be present on the item and should be washed out. This can be accomplished most easily with a good modern detergent, but *not* one containing whiteners and/or brighteners. I usually use Ivory Liquid or Tide. Wash and squeeze the material a couple of times and rinse well. Cellulosic materials may then be soaked for several hours, dried, and used, but protein fibers should always be given an after-rinse of 15 to 60 minutes with weak sulfuric or acetic acid or vinegar. The acidic after-rinse neutralizes any remaining alkali which would be detrimental to protein fibers, and it often improves the color to true indigo. It may also decrease fading; indigo does fade somewhat, particularly in pale shades. It also restores the luster and sheen to silk. I have gotten into the habit of using the after-rinse for all items, though neither acetic acid nor vinegar is cheap. One plastic teaspoonful (1 ml) of commercial 40-99 percent sulfuric or acetic acid to the gallon of water is plenty. One or two ounces of vinegar, which is only 5 percent acetic acid per gallon, will do. Again, the after-rinse is much more important with protein fibers; remember what effect this has on the hair, after washing. Acetic acid may be obtained from chemical or photographic supply houses, or ordered through your druggist. Soak the item in clear water (1 to 24 hours) following the acid rinse, squeeze and dry. This long final rinse may be necessary to remove all traces of reducing agent which might eventually produce an uneven bleaching effect.

Trouble-shooting the vat:

• If the vat color changes from yellow or yellow-green to bluish green, and its ability to dye decreases, this indicates that the indigo is changing to the blue oxidized state and that the vat requires more lye, reducing agent, or both. If a stronger vat (more indigo) is desired at this point, both reducing agent and lye will be added with the addition of more concentrate which will probably correct the problem.

If a stronger vat is not required, take up two small white styro-foam cupfuls from the vat. Add a little lye to one, and a little re-ducing agent to the other and stir very gently. This should indicate which is needed, or both if neither cupful resorts back to the yellow-green color. Add either or both to the vat in small amounts, being particularly careful with the lye since excess lye is a far greater problem than excess reducing agent. In fact, many indigo dyers al-ways use quite an excess of reducing agent, although in great excess it causes problems as well. For example, if way too much reducing agent is present, the indigo will not be as readily taken up by the fiber and it will not oxidize fast enough upon airing, allowing greater color development at the ends of the skeins where gravity pulls the liquid (Matthews, 447). If the vat lacks sufficient reducing agent, too much of the indigo in the vat will be unreduced; the material may not be well penetrated by soluble reduced indigo, and the color will develop too rapidly upon exposure to the air, mostly on the sur-face. Complete visual color change to blue should take 30 to 60 seconds.

Too much lye prevents adequate absorption of indigo by the fiber (the indigo white is too soluble), and it is almost impossible to build shade by redipping (particularly on cotton and linen) because about as much previously deposited indigo on the fiber is re-reduced and stripped as is added. Insufficient lye or other alkali manifests itself by uneven dyeing or spotting. Apparently, the ideal pH for this type of vat is 8.3, just where colorless phenolphthalein turns to pink. In laboratory tests with a pH meter, I have gotten good results from pH 7.5–9.0.

In the old days, the dyer would judge degree of alkalinity both by feel and taste (fermentation vats). A vat that is pH 9 or above feels slippery, already at or beyond the optimum alkalinity level for this type of vat; pH 7 is neutral; pH less than 7 is sour to the taste; pH above 7 is bitter to the taste. But these people were experts and specialists; neither method is recommended now.

• If the dyeing turns out to be the correct shade, or slightly less, but the color is not level, there are several possible causes. Unlevel dyeing can result from improper scouring; an unbalanced vat (see below) in which the material is too crowded in the pot or lying too long crumpled up on the bottom; dyeing an old unevenly worn item, some areas of which take the dye more readily than others, or dyeing a new item of nonuniform texture. If the vat is in good order, rein-troduce the wet item (uncrowded), and move it about gently for about

5 minutes. This may level the color. Give the item an additional dip if the first one effects an improvement.

• If the color is level (or unlevel), but the shade is too dark, set up a new vat, containing water, alkali, and reducing agent, but *no indigo*. After waiting 15 minutes or so for reduction, introduce and work the wet item for a few minutes. Some of the indigo on the fiber will be reduced and stripped into the vat. This not only reduces the shade on the item but usually has a leveling effect as well. The method works far better with cotton than silk or wool. For large unwieldy piece goods or skeins, unlevel dyeing can sometimes be corrected by placing the item in a large tub of water immediately upon removal from the vat. The item will oxidize from the oxygen in the water. Agitate the item once it is in the water. This was commonly done in the old days when, say, a chain of between 10 and 15 hanks of yarn, 2 to 3 pounds each, were taken from the vat. There was no way to squeeze them all out evenly before oxidation took place.

• If the job is unacceptable and the decision is made to strip the color entirely, prepare a vat as above, add the article and heat to 160° to 190° F for 10 to 30 minutes. Indigo in the reduced state is destroyed by temperatures in excess of 145° F or so. Incidentally, Rit color remover is a combination of alkali (sodium carbonate) and reducing agent (sodium hydrosulfite). Such color remover will not strip all colors, but it will destroy all indigo if the material is left in the vat long enough, particularly cotton and linen. (If the item is silk, however, the chances are that most of the sheen will be destroyed permanently.) After stripping, the item should be washed thoroughly, and put through the sour bath.

Building to shade or building a series of shades:

Let us assume that the shade is not dark enough for your project, or that you wish a series of shades. Divide the material according to the number of shades desired and proceed as follows: Check the color of the dyebath by looking, directly or by taking up a cupful in a white container. Is it still yellowish green? If not, follow the directions above. Wet the material thoroughly and reintroduce it into the vat (all pieces or skeins for a series of shades). Agitate slightly and watch the color of the material. If it remains blue or changes color to no more than green, the chances are that a heavier shade will develop. On the other hand, if the color changes too much toward the yellow, it is unlikely that a heavier shade will be produced, and the material should be removed from the vat and again allowed to air. The vat is probably too alkaline or contains too much reducing agent and color is being stripped and added at the same time, as previously described, making building a deeper shade more difficult. If such is the case it would be well to take a coffee or tea break (30 to 60 minutes). During this interval the vat will become less alkaline because of absorption of carbon dioxide from the air, which in solution produces carbonic acid, and/or the amount of reducing agent will decrease. The carbonic acid helps neutralize some of the alkalinity produced by the lye. Shorter dips may also be helpful (maximum of 5 minutes). Leave the material in the vat only until it turns green but not yellow, remove and air. Ideally, the material will remain blue during redipping. If the dyed material does not become yellow during redipping, and still does not become darker in color, more indigo concentrate should be added to the vat. When building shade, it is best to have only enough reducing agent and lye—particularly lye—to keep the vat in order since less already-deposited indigo on the fiber will be stripped. Redip those items wanted darker until the series is built.

Additional notes on the lye-hydrosulfite vat:

• This vat is ideal for light to medium shades on cotton, linen, and silk, and for all shades from the lightest to the darkest on wool. Heavy shades on cotton, linen, and silk are possible using concentrates containing 4 or more teaspoons indigo paste per quart.

• The same strength vat which will produce light to medium shades on cotton, linen, and silk will produce a much darker shade on wool.

• If the vat is strong, do not dye woolen items at high temperature, i.e., 120° to 140° F, because the wool will take up the indigo too fast and perhaps produce too dark a shade. Also, if the indigo is taken up too rapidly, it is likely that it will crock. Reduced indigo behaves as a very weak acid that is rapidly attracted to the alkaline character of wool, a condition that does not exist with the cellulosic fibers or with silk. The speed with which the indigo is taken up by wool decreases at lower temperature. Interestingly enough, indigo is more rapidly taken up by cellulose at lower temperatures (90° to 110° F). I suggest, then, these lower temperatures be used if both cotton and wool are being dyed in the same vat. In this case, dye the cotton first.

• If a woolen item crocks badly, it will probably continue to do so. If this is the result of greasy yarn, the yarn should be scoured well with detergent. If the crocking is the result of dyeing at too high a temperature or in too strong a vat, the item should be treated in a bath containing water, lye, and reducing agent only (previously described). This treatment will lighten the color, and should redeposit the indigo in noncrocking condition. The item should be in the bath 20 to 30 minutes.

• Building shade takes time, and is sometimes done over a two-day period, particularly with cotton or linen. This vat will dye well until its temperature drops to 80° F. If the vat is to be left overnight and used the next day, add 1 to 2 teaspoons of thiourea dioxide or 3 to 6 teaspoons of sodium hydrosulfite, depending upon the size of the vat, and 1 teaspoon of lye. Stir gently, and cover if possible. The vat will be in good order for dyeing the next morning if its temperature is 80° F or higher. If not, reheat the entire vat if possible, or remove a portion and heat to 140° F and carefully add this portion back to the vat. If this procedure raises the temperature of the entire vat above 80° F, dyeing may commence an hour or so thereafter. This procedure may be used anytime the vat temperature falls below 80° F. For rejuvenation of an idle vat, heat and add lye, reducing agent, and detergent.

Advantages of the lye-hydrosulfite vat:

• It is the easiest and fastest of the indigo vats to set into order, a procedure that takes only one hour or so.
• There is usually no sediment; all other vats have sediment which the material being dyed should not touch.
• It is the easiest vat to use with very large skeins or large pieces.
• There is less loss of indigo compared to other vats, only 1 to 2 percent (Matthews, 429).
• It is the best vat for wool, in most cases, and if used at a temperature of 80° to 110° F, and with minimum alkalinity, will not shrink or felt wool. This is an important consideration when dyeing or redyeing a woolen sweater, for example.

Disadvantages of the lye-hydrosulfite vat:

• It is difficult to build very heavy shades on cotton, linen, and silk unless the vat contains a large amount of indigo, or unless you are willing to give the item many dips (10 to 20 or more). Adding the material dry helps.

- Building shade on cotton, linen, and silk is more difficult than with the zinc-lime vat because thiourea dioxide or sodium hydrosulfite are strong, fast-acting, reducing agents. Thus, the zinc-lime vat is easier for shade building because it normally has a very high indigo content and uses a very slow-acting reducing agent (zinc dust). It takes zinc about 4 to 8 hours to reduce the indigo in the vat, even though the vat is very alkaline. Therefore, when you re-dip cloth in this vat, very little of the already-deposited indigo oxidized on the fiber will have time (in say a 3 to 20 minute dip) to be re-reduced by the zinc and stripped, while reduced indigo in the vat will be added. In the lye–thiourea dioxide or sodium hydrosulfite vat, shade cannot be built readily unless the vat is only minimally alkaline and contains minimal amounts of reducing agent, because the reducing agent is so fast acting and thus can strip already-deposited blue indigo. Again, the problem of building shade with indigo on wool is much less than with cotton, linen, and silk because of the great chemical attraction between reduced indigo and wool.

Recipe No. 2: The Zinc-Lime Vat

This type of vat has been in use since about 1845. It is considered by many to be the best vat for medium blue to almost black shades on cotton, linen, and silk, and is usually the vat of choice for *adiré, plangi, shibori* (tie, fold, stitched resist work). It is sometimes referred to as the "Japanese kimono vat." It should not be used for wool because of the high alkalinity (pH 10.5–12.5).

The vat uses lime (calcium oxide, CaO: old fashioned "privy" lime) for the alkali and zinc dust for the reducing agent. The essential chemistry is:

$$CaO + H_2O \rightarrow Ca(OH)_2$$
$$Ca(OH)_2 + Zn \rightarrow ZnO_2Ca + H_2.$$

The hydrogen removes the oxygen from the vat by combining with it, forming water; it also reduces the indigo blue to indigo white.

Zinc dust and calcium oxide (calx) may be obtained from suppliers such as Cerulean Blue. Zinc dust may also be obtained from chemical supply companies. Purchase only that which is very finely ground into a dust. Calcium oxide is also marketed as slaked lime or quicklime and sold in building material supply stores. It may also be purchased at garden supply stores. Unless considerable good quality natural indigo is at hand, it is strongly suggested that the much less expensive finely ground synthetic powder be used.

This vat may be kept in continuous or intermittent use for months (with not too much attention) and will dye at temperatures between 60° and 115° F. Therefore, a vat can be set and, with only a little attention, used as needed. It can also be rejuvenated and brought back into use after several months' rest.

The basic instructions for setting the vat have remained the same for many years. As with Recipe 1, the first step is the production of a stock solution which will be added to the actual vat. Since a sediment is present in this vat which should not touch the material being dyed, it is best set in a deep, relatively narrow container. My favorite is a 16-gallon white plastic container, 27 inches high by 12 inches in diameter, with a lid. Narrow, deep plastic garbage cans with lids work very well. The recipe given will be for 15 to 20 gallons; increase or decrease the amounts proportionally for larger or smaller vessels.

Preparation of the stock solution:

(Caution: If you are working indoors it would be wise to put on a painter's face mask rather than chance inhaling finely ground zinc dust, lime, or indigo.)

1. Pour 4 to 8 ounces of boiling water into a previously heated pint or quart mason jar and quickly add, with stirring, 1 3/4 ounces (50 grams or 9 level tablespoons) of finely ground indigo powder. Place the lid on the jar so that the hot water and steam wet the indigo thoroughly. This is the indigo paste. Keep hot in a pan of heated water (120° to 160° F).

2. Mix 1 to 1 1/2 ounces (2 1/2 to 3 1/2 teaspoons) of zinc dust with about 8 ounces (1 cup) of hot water in a separate stock vessel of 2-quart to 1-gallon capacity (glass canning jars or other heat-resistant glass jars with a lid work best).

3. Gradually add the indigo paste to the zinc dust solution, stirring.

4. In another vessel put one pint of hot water and add 4 ounces (1 cup) of lime. Stir well for 2 or 3 minutes or until a relatively thick creamy paste results.

5. Add the lime to the stock vessel, stir well, and place the lid on the jug.

6. The stock vessel containing all ingredients should be kept as close as possible to 120° to 140° F — at least for the first hour. This may be accomplished by placing the jug in a large vessel of hot water.

7. The stock should be stirred occasionally, especially during the first hour.

8. When reduced, the stock solution will be dark amber brown or greenish yellow brown with a sediment. Reduction usually takes 3 to 5 hours. The stock may be added to the prepared vat at this time, or it may be kept for one or two weeks.

Preparation of the dye vat:

1. Place 15 to 19 gallons of water at a temperature of at least 65° F but not exceeding 140° F in the plastic garbage can or other similar container. Hot water from the tap is fine.

2. Make a paste with 1 1/2 ounces (4 tablespoons) of lime and enough water to make a creamy consistency.

3. Add the lime to the vat and stir well with a dowel rod or, better yet, a one-by-three board or similar paddle.

4. Add 1/2 ounce (1 to 1 1/2 teaspoons) of zinc dust and stir well again.

5. Permit at least 30 minutes to elapse to allow time for the vat to be rendered in a reducing condition before adding the stock.

6. Before you add the stock solution, a decision needs to be made. If all of the stock is added, dyeing will produce quite dark shades, very quickly. Such a vat will be quite suitable for tie-dyed work, where extreme contrast – achieved with 3- to 5-minute dips – is desired. But if lighter shades or a group of shades to be produced by redipping are desired, add only half of the stock.

7. Gently stir the stock and add half or all of it by lowering the jar into the vat before pouring out. This, of course, avoids splashing and introduction of air.

8. Stir the vat again, cover, and allow to stand until the fluid of the vat is a clear yellow (beer colored) or dark amber (bock beer). The clear yellow will be the weaker vat. This interval will likely be 5 to 24 hours. A heavy bluish purple bubbly froth or scum (the "flower") will form on the surface of the vat when the vat is in good order. This froth must be pushed aside when you are judging whether the vat is reduced and in good order. After pushing aside the flower, it is easier to judge color and vat condition by taking up a white styrofoam cupful of the fluid. This vat always contains considerably more unreduced indigo than does the lye-hydrosulfite vat. Therefore, it should not be used when the fluid is green since then there will be even more unreduced indigo to wash out of the dyed items. An

excess of zinc may keep the vat "muddy" because of rising hydrogen. If such is the case, stir the vat and wait until the liquor is clear.

Dyeing procedure:

1. Take a paper plate and carefully remove all of the flower, which contains large amounts of unreduced indigo blue. Save this material in a separate vessel, since it should be put back into the vat when the dyeing session is over.

2. Put on rubber gloves and introduce the well wetted-out or dry and scoured material (tie-dyed materials may be added dry). The material (skeins or piece goods) must be put in by hand or placed on a dowel rod or bent iron rod or on a stretcher since the material should not touch the sediment in about the bottom one-fourth of the vat. The other alternative is to introduce the material into the vat in a well wetted-out net. It is necessary to move the material about somewhat to obtain level dyeing. Skeins may be turned over the rod so that all parts are under the surface the same length of time or the same depth in the vat. A rather short first dip of about 5 to 10 minutes is suggested; this will be long enough to give some idea of the strength of the vat.

3. After squeezing the item out underneath the surface, remove it from the vat and give it the air. An item should be aired for at least 20 minutes before redipping, and 30 is better. If the item is silk it should be placed in water if it dries noticeably before redipping.

4. After airing, the item may appear quite dark even though it was in the vat only five minutes. Items dyed in this vat appear even darker when wet than those dyed in the hydrosulfite vat because of the larger amount of unreduced indigo present on the material. Regardless of the color, at least one additional dip is recommended because of the very short time the material was in the vat during the first dip.

5. When the correct shade has been built by redipping, cotton or linen should receive its final airing of an hour to 24 hours. Silk should be placed in water when it starts to dry.

6. After airing, the material needs a couple of good washes in detergent, either by hand or in the washing machine. Following this, a sour rinse of weak acetic acid (see Recipe 1) should follow for all items, including cotton and linen, to remove the calcium from the lime water and to remove any residual alkali. Follow this with a clear water rinse and dry.

Additional notes on the zinc-lime vat:

- Items appearing to be dyed nearly black with this vat, when wet, will be dark blue when finally washed out and dry.
- Hard cotton and linen yarns should be dyed in a relatively weak vat, with several dips, to insure even color and good penetration. Some dyers keep a strong and a weak vat on hand. Adding the item dry rather than damp also facilitates penetration.
- Shades may be built quite easily by redipping.

Maintenance of the zinc-lime vat:

- At the end of a dyeing session pour the "flower" back into the vat and stir up the vat from the bottom so as to mix the sediment thoroughly with the vat contents. Most dyers then stir the vat in circular direction for a few seconds to concentrate the flower in the central part of the vat.
- After it has been stirred, the vat should be in order for dyeing again in about two hours. Do not resume dyeing until the vat is clear. Stirring undoubtedly serves to help release hydrogen trapped in the sediment and to accelerate reaction between the zinc and lime water. The vat will remain in order longer if it is stirred once a day whether it is used or not.
- No further additions will probably be necessary for at least 2 days to one week unless quite a bit of dyeing is being done.
- Eventually the flower will begin to fall, and when this happens the vat must be "sharpened." The flower will lose its purplish glistening bubbles and perhaps become a bluish gray in color. It will also fall or appear flat. In addition, the vat color may change from brown or yellow to a greenish yellow. If dyeing is to be done every few days, experienced dyers judge the condition of the vat on a daily basis and make addition of lime and zinc as needed. This requires some experience. It is suggested that one tablespoonful of lime be added to a cupful of vat fluid, stirred, and then 1/3 teaspoon of zinc dust added to the same cupful of material and the mix stirred again. Add this to the vat and stir the vat. Resume dyeing when the vat is restored (well-developed flower and color).
- If a lot of dyeing was done on the first and/or second days, lime at least will need to be added, and zinc and indigo may be necessary also. At any rate, additional indigo will be necessary eventually when the vat becomes weaker than desired. The general rule is to add a

runny, pasted mix of indigo-lime and zinc at a 5:10:3 ratio – for example 5 grams of indigo, 10 grams of lime, and 3 grams of zinc dust. (Five grams of indigo is about 1 level tablespoon, 10 grams lime, one level tablespoon, and 3 grams zinc, 1/4 teaspoon.) Of course, a completely new, completely reduced stock, or part of a reduced stock may be added also.

   • Some recipes suggest addition of more lime, more often, than is suggested here. I try to avoid this because of the extremely high alkalinity it produces. Excess lime will probably raise the vat pH to around 12. If this occurs, it is especially important to wash out all residual lime and carefully use the sour bath in finishing. Too much zinc will keep the vat in a "muddy" condition since the rising hydrogen, if in excess, moves sediment from the bottom into the vat fluid. If this occurs, suspend dyeing until the chemical reaction slows down (often within 24 hours) and the vat fluid becomes clear.

   • If the dyed item turns out too dark, after washing decrease the shade (see trouble-shooting section for Recipe No. 1). A lighter shade can also be obtained by shorter dips, but penetration may be incomplete, resulting in a surface effect only. The color may be reduced somewhat by oxidation under water. For this process, quickly move the material from the dyebath to a pail of water. Agitate and then leave in the water for 30 minutes or so.

Rejuvenation of an idle vat:

This type of vat may be rejuvenated even after several months' rest. A relatively fast way is as follows:

   1. Remove the crust and crystalline material (calcium carbonate) from the surface of the vat.

   2. Stir the vat thoroughly and then remove several gallons to a vessel that may be heated. I generally remove and heat 5 gallons from a 15-gallon vat to 140° F, and then add this back to the vat. The vat temperature is thus raised to 90–100° F.

   3. Now add new stock, or a slurry of zinc and lime if the vat already contains sufficient indigo. If only the zinc-lime mixture is added, keep the ratio at 3 grams of zinc to 10 grams of lime (1/4 teaspoon of zinc to 1 tablespoon of lime). Try a slurry of 1 1/2 teaspoons of zinc and 5 tablespoons of lime. This may or may not be sufficient. If it is, the vat will be restored to dyeing condition within a day or two. Stir the vat thoroughly morning and night. Remember that the "flower" should reappear and the vat color should change to yellow-brown.

Advantages of the zinc-lime vat:

- It is easy to build a series of shades, including very dark shades, using this vat on cotton, linen, and silk.
- Because it builds dark shades quickly, it is ideal for any tie, fold, resist, or stitch dyeing.
- It may be set up and used intermittently with little attention.
- It will dye until the vat temperature drops below 60° F.

Disadvantages of the zinc-lime vat:

- It has a sediment in the bottom which the items being dyed should not touch.
- It is very alkaline, making a sour rinse of the dyed items mandatory to remove alkalinity, calcium, and many times to restore desirable color. A very alkaline vat gives a bluish gray color to indigo. The sour rinse is also necessary to restore some of the luster and sheen to silk.
- Because of the very high alkalinity, woolen items should rarely, if ever, by dyed in this vat.
- Loss of indigo in this vat due to secondary chemical reactions is about 10 percent (Matthews, 429).

Recipe No. 3: The Copperas or Ferrous Sulfate Vat

This vat is suitable only for cellulosic materials (cotton and linen) because the reducing agent is an iron compound which is quite readily taken up by protein fibers (wool and silk).

The method originated abut 1750 and lasted until about 1914. Recipes for the method appeared in dye manuals as early as 1789 (Hellot/Macquer, 451). It is quite possible that the method was used even earlier in India, but I have no definite proof. Until largely replaced by the zinc-lime vat, the ferrous sulfate vat was in most cases a much faster and more efficient method for cotton and linen than the fermentation vat.

The method depends upon lime or lime and soda ash (sodium carbonate) or pearlash (potassium carbonate) for the alkali and copperas (ferrous sulfate) for the reducing agent. The essential chemistry follows (Matthews, 422–23):

$$FeSO_4 + Ca(OH)_2 \rightarrow Fe(OH)_2 + CaSO_4$$
$$2Fe(OH)_2 + 2H_2O \rightarrow Fe_2(OH)_6 + H_2.$$

The hydrogen removes oxygen from the vat by combining with it and forming water; the hydrogen also reduces the indigo blue to indigo white.

The vat may be set up rather quickly and simply, with or without a stock solution. Copperas is a faster-acting reducing agent than is zinc dust (see Recipe No. 2: The Zinc-Lime Vat), but the loss of indigo due to secondary chemical reactions is greater (about 25 percent), and the amount of sediment is greater. In fact, once the vat is set, only the top two-thirds, or a little less, is suitable for dyeing. A very deep, narrow vessel is therefore required. Two recipes are listed for the copperas method—one made directly, the other with a stock solution.

Direct method:

The recipe is essentially that of Mairet (48); similar recipes are in Bemis, Bronson, and other nineteenth-century dyebooks. The quantities listed are for a small experimental vat of 5 to 6 gallons, useful for small items and for determining whether you wish to set a larger vat. Increase all materials proportionally for a larger vat. I have had excellent success with this vat, and it usually comes into order in one day or less.

1. Make a paste with 2 ounces (10 tablespoons) of synthetic indigo or 3 ounces (15 tablespoons) of natural indigo and hot water as described in Recipe No. 1. (Synthetic indigo is suggested because of the lower price.)

2. Dissolve 8 ounces (20 tablespoons) of copperas in one quart of hot tap water. Good quality blue-green copperas works best.

3. Make a runny slurry of 10 ounces (30 tablespoons) of lime with about 1 quart of hot tap water.

4. Nearly fill a tall, narrow plastic, glass, porcelain-lined, or stainless vessel with 5 to 6 gallons of room-temperature water.

5. Add, stirring, the indigo, copperas, and lime in that order.

6. Stir well for two or three minutes, cover, and let stand until the vat is in order. In proper condition the vat liquor should be clear brownish yellow with blue scum on the surface. Bluish veins may also run through the fluid. If much bluish scum "flower" is present, it should be removed and saved as with the zinc-lime vat prior to dyeing.

Dyeing procedure:

The dyeing procedure is exactly the same as for the zinc-lime vat except that the sediment, which the fiber should not touch, makes up a greater percentage of the vat.

Reproduction of this method of dyeing in Scheesel, Germany (nineteenth-century method), as reported by Blumrich (1984, 23), involved ten 20-minute dips with 20 minutes' airing between each dip for dark blue resist-dyed cotton fabric. Fabric so dyed would remain a permanent blue as long as it lasted. This number of dips is not essential unless a deep blue is desired, but the 20-minute dips with 20 minutes' airing in between will result in the desired shade. If possible, skeins so dyed should be lowered into the bath on bent iron rods so that the entire skein is beneath the surface, and the skeins should be turned at intervals.

Maintenance of the vat:

Because of the continual loss of indigo due to secondary chemical reactions, it is suggested that the vat be used rather often over a relatively short period, say 2 weeks, until exhausted, and then new vats set. Often the dyer of old would have 4 or 5 vats on hand, the newest vat being the strongest. In building shade, the first dip would be in the weakest (oldest) vat, the second dip in the next oldest (slightly stronger), and so on until the desired shade was achieved. Such vats could be kept in order by additions of copperas or lime or both. Again, color of the vat fluid is the best sign. Copperas should be added (1/2 ounce, 1 1/2 teaspoons) when the color changes from brownish yellow to green, and the vat should then be stirred well. Lime (1/2 ounce, 1 1/2 tablespoons) should be added if the color changes from brownish yellow to a bluish cast. The flower should be added back to the vat and the vat stirred at the end of each dyeing session. It is best to stir the vat each day whether it is used or not.

As with the zinc-lime vat, additions of too much lime and copperas should be avoided. Too much lime retards the attachment of indigo to the fiber; too much copperas does much the same, producing a weak color.

The ideal temperature of the vat is about 70° F.

Stock solution method:

This recipe is for a 15- to 20-gallon vat. Increase or decrease the materials proportionally for a larger or smaller vat. The recipe is essentially that from *Manual for the Dyeing of Cotton and other Vegetable Fibres* (217).

Preparation of the stock solution:

1. Paste 5 to 6 ounces of synthetic indigo or 10 to 12 ounces of natural indigo with hot water (about 120° F).
2. Dissolve 3 pounds of copperas in sufficient hot water to dissolve.
3. Paste 3 1/2 to 4 pounds of lime also in water at about 120° F.
4. Pour all three materials, stirring, into a vessel with a capacity of about 4 gallons and fill with water at 115° to 120° F.
5. Permit the stock to stand until reduction is complete (vat liquor brownish yellow in color); this should take 2 to 3 hours. (Note: This stock should not be kept more than 2 or 3 days after it comes into order because of excessive loss of indigo due to secondary chemical reactions.)

Preparation of the dye vat:

1. Make a paste of 8 ounces lime with hot water.
2. Dissolve 2 to 3 ounces copperas in hot water.
3. Partially fill the vat with 10 to 15 gallons of water (65° to 100° F).
4. Add the lime and copperas in order, stirring.
5. Allow one hour to pass before adding all or part of the stock.
6. The vat will probably be in order 2 to 3 hours after adding the stock (blue flower, brownish yellow vat fluid).
   - The dyeing procedure is the same as with the direct vat.

Advantages of the copperas vat:

- The vat is quickly set into order in that copperas is a faster-acting reducing agent than is zinc dust (though slower-acting than sodium hydrosulfite or thiourea dioxide).
- It is somewhat easier to keep in order than a zinc-lime vat.
- The vat can produce heavy shades rather easily by redipping and thus is quite suitable for resist work and building shade.
- The working materials (copperas and lime) are quite readily available (farmer co-op and garden supply stores).
- The vat works quite well even at 70° F or slightly less.
- The vat may be set up directly, rather than with a stock solution.

Disadvantages of the copperas vat:

- A lot of indigo is required in that about 25 percent is lost because of secondary chemical reactions.
- A very deep, narrow vat is necessary because of the large amount of sediment.
- The vat should be used and exhausted within 2 or 3 weeks after being set up because of the continual loss of indigo.
- The vat is suitable only for cotton, linen, and other cellulosic materials.

Fermentation Vats

These vats are the oldest and were in use at least from 2000 B.C. to, in a few instances, the present (traditional dyers of Japan, possibly isolated areas in Nigeria, and some contemporary dyers still use fermentation vats). Their use decreased markedly for cotton and linen when the copperas vat became available (1750), and for wool when the hydrosulfite vat became available (about 1880).

All natural fibers may be dyed in fermentation vats, but they are generally more suitable for the dyeing of wool than for linen, silk, and cotton. One reason is that they must be warm to work well, and another is that they are generally weaker with respect to reduced indigo. The cotton dyer of old, especially, had to rely on many dips, up to fifty in some cases over several days, to obtain a very deep blue. Of course, with such a large number of dips, the color would be very, very permanent. The weaker vat was more difficult and time-consuming for the various forms of tie, stitch, or fold dyed work as well.

Fermentation vats rely on bacterial or yeast fermentation to reduce the vat (remove oxygen and reduce the indigo blue to soluble indigo white), and ammonium hydroxide (ammonia) and ammonium carbonate or lime water or wood ash lye (weak potassium hydroxide and carbonate) to dissolve the reduced indigo. In virtually all cases, alkali and reducing agent are minimal in the working vat. Thus, the worry of excess alkali or reducing agent is minimal.

Fermentation vats, though somewhat slow and requiring considerable expertise and attention, were responsible for *all* of the fine indigo blue found on existing wool specimens dyed prior to about 1880 and much later in certain places. They were eventually replaced by the hydrosulfite vat. Fermentation vats were also used some places rather late for cotton and linen as well (Nigeria, etc.) but were mostly replaced by the copperas vat (1750) or the zinc-lime vat (1845).

All fermentation vats have a smell, the urine vats generally being the worst. Thus, it is suggested that they be set up and used during summertime when they may be kept outside in the sun. Under these circumstances the proper temperature may be more easily maintained and the smell kept outside.

The alkali of the urine vat (ammonia–ammonium hydroxide and ammonium carbonate) is the least damaging of all alkaline substances to wool, and for this reason buyers of indigo-dyed woolen yarns and woolen articles often would not purchase such items if they did not smell right (the stale urine odor reminiscent of an old-fashioned diaper pail or privy does disappear eventually). This attitude prevailed in some places long after the introduction of the hydrosulfite vat. And it is dismaying to find that cheating occurred everywhere, as I recently discovered in the following:

> The typical indigo smell which adheres to all dyeings produced on the fermentation vat, may also be produced artificially by the addition of Vat Odour N. This is a volatile perfume which is applied by spraying the finished goods with the solution, or by turning the dyed goods in a cold bath containing about 1 1/2 ounces of Vat Odour N, dissolved in 2 pints alcohol, in 100 gallons water. The preparation may also be added to the finishing paste. Subsequently the goods so treated are dried not too quickly. (*Manual for the Dyeing of Cotton and other Vegetable Fibres*, 221)

Note that no mention was made concerning the composition of Vat Odour N.

The question may arise as to why the craftsperson should learn to use the fermentation vat at all when the lye-hydrosulfite vats or the zinc-lime vats are much more easily set and managed. Five reasons come to mind: some shades are obtainable only with a fermentation vat; materials such as wool fleece and cotton lint may be left for extended periods (several hours to 1 to 2 days) in such a vat, generally producing a fine and quite permanent color; once working, a fermentation vat may operate for several days with little attention; building shade (over time) generally works well in the fermentation vat because the amounts of alkali and reducing agent are generally optimal; and, finally, it is gratifying to be able to reproduce a method having such historical significance. As previously stated, all existing museum specimens dating from at least 2000 B.C. to 1750, and much later in many cases, were done in fermentation vats.

Perhaps the urine vat is the easiest fermentation vat to start with (on a small scale) because it often works with no additions whatsoever except indigo and urine. Urine contains nutrients for bacterial fermentation which reduces the indigo, and the bacteria also convert the nitrogenous waste product (urea) into ammonia (ammonium hydroxide and carbonate), the alkali which dissolves the reduced indigo white.

Which type of urine is best? This was the subject of considerable debate in the old days, with many "experts" contending that urine from diabetics and drunkards was generally best. This conclusion has scientific credibility; urine from untreated diabetics is high in urea and volume, and contains blood sugar (glucose or dextrose), an additional nutrient for bacterial or yeast growth. Alcohol and its breakdown products can also serve as nutrients for microorganisms. Urine from small boys and pregnant women was also considered very good. In the case of pregnant women, at least, there are probably additional nutrients and urea produced by the developing fetus.

Actually, any urine, such as is produced by normal eating and drinking habits, will do. This will amount to from 1/2 to 1 1/2 quarts per day. It is unsound to embark on a high liquid intake in order to produce the desired quantity of urine faster. This idea is an old one. I recently read an account of a nineteenth-century English woman who would invite all of her husband's men friends over for an evening of card playing and strong cider drinking whenever she wished to set a reasonably large urine vat. This resulted in some amusement, with neighbors commenting that "Madame so and so" was having another of her "piddle parties." The reason the method is unsound is that the adult body produces and releases into the urine approximately 3 grams (1/9 ounce) of urea per day whether this amount be released in 1/2 to 1 1/2 quarts of urine or in 3 to 4 quarts of urine. Therefore, the nutrient concentration will be low per unit volume in the dilute urine.

It is suggested, however, that vitamin and mineral intake be in creased several days prior to collection in that excesses of these materials (B complex vitamins, vitamin C, and minerals) are excreted in the urine. Increasing protein intake one or two days prior to and during the collection period will increase both urea production and volume of urine, thus shortening the collection period as well. The reason for this is that excess ingested protein is not stored in the body; instead the excess amino acids from the protein are deaminated (the amino group is removed and converted to urea), and increased urea removal by the kidneys is accompanied by additional water loss.

It is also strongly suggested that some thought be given as to where to store the urine during the collection period. My first experience proved somewhat embarrassing from this lack of foresight. I had collected about two-thirds of a gallon in a white plastic milk jug, and hoping to accelerate the transformation of some of the urea to ammonia, set the jug out in the sun next to the garage. Somewhile later, one of my sons decided to cut the grass, and upon finding an empty gas tank, reached for the nearest source of what appeared to be gasoline. He poured at least a quart of the liquid into the tank before the odor reached his nostrils!

Sig (Urine) Vat No. 1 (Sig Vat, "Good Old Sig Vat," Chamberley Vat)

It is suggested that, initially a 1- to 3-gallon vat be attempted since this may be set up in a strong wide-mouth glass jar or white plastic container (to better observe the color change from blue to green or yellow-green as the vat comes to order).

1. Collect the desired volume of urine.

2. Make a little bag of the finest mesh material on hand and place 1/2 to 1 ounce of *natural* indigo (2 to 6 tablespoons) within. Tie the bag shut and drop it into the vessel or suspend it just below the surface tied by a string to a stick laid across the top of the vessel.

3. Place the vessel in a warm place or in the sun (in the old days, in cold weather, the vessel was placed on the hearth next to the fire). Natural fermentation should start within a couple of days. Also, night and morning rub the bag of indigo between the fingers to release into the vat the indigo blue. Eventually, all of the indigo fine enough to reduce will be released.

4. If all goes well, in a few days, depending largely upon temperature, the vat color will change from blue to green or greenish yellow, at which time it is fit to use.

Dyeing procedure:

1. If little sediment is present, well scoured and wetted-out articles may be placed directly in the vat; otherwise they should be suspended or dipped. An exception is wool fleece or cotton lint which may be introduced, uncrowded, in a plastic mesh bag, such as those used for onions or oranges. Immersion time can be a few minutes to several hours to 1 or 2 days. Air the material as for other indigo vats, redip for heavier shades, and finish off as with the lye-hydrosulfite vat.

2. Watch the color of the vat carefully and add no more air than absolutely necessary when adding or removing materials from the vat.

3. Add more urine as evaporation occurs. This replenishes both nutrients and urea, which the ferments use up. Also add more indigo as time passes. This vat may be used for some time.

A fermentation vat is a "chemical balancing act." On the one hand, the fermenting microorganisms produce the alkali (ammonia) from the urea, but at the same time they produce carbonic acid and reduce the indigo. Carbonic acid partially neutralizes the alkali, producing very weakly alkaline ammonium carbonate. Thus, if fermentation is too active, the indigo will be reduced, but there may not be enough alkali to dissolve it; i.e., the vat may remain blue and go sour. Of course, if fermentation is not active enough, little of the indigo blue will be reduced to indigo white.

Trouble-shooting the vat:

• If the vat remains blue after several days have elapsed, the temperature may not be high enough to induce fermentation. Get the vat into the sun or into a warmer area. Add a little dry cake yeast, madder root, or dried woad, as the vat may not have become "seeded" with the proper microorganisms.

• If fermentation has been heavy and the vat smells sour, add, cautiously, small amounts of clear liquid household ammonia. The alkali will decrease the growth rate (fermentation) of the microorganisms, supply the necessary alkali for the reduced indigo, and at the same time more than neutralize the acidity. In many instances urines contain approximately the appropriate amounts of nutrients and urea so that enough ammonia is produced to keep fermentation at the level of reducing the indigo, but not rapid enough to sour the vat.

• Some indigoes, particularly synthetic indigoes, will not work in a fermentation vat. Apparently this is the result of chemicals used in the processing or production of the indigo which prevents fermentation.

• If fermentation seems weak, add a small amount of Karo syrup. Even table sugar is all right, but add these materials cautiously, say 1 teaspoon at a time.

Sig (Urine) Vat No. 2: Modified Vat

Often in the past materials were added to the urine in order to get the vat into order more quickly. Recipes involving additions are found in Bronson (184); Bemis (154–55); Grae (59-60); Hummel (308); and Kuder (187), to name a few. Some call for the addition of about 1 ounce of powdered alum or common table salt in 5 gallons; others call for the addition of both alkali and an additional sugar source such as dates, raisins, madder, bran, or molasses. The following is a slightly modified version of that listed in Bronson (184) and will probably dye 2 to 5 pounds of wool a medium to heavy blue and the same quantity of cotton or linen a medium blue (with sufficient dips).

1. Collect a pailful of "chamberlye" (about 3 to 5 gallons) and place this in the vessel to be used for a vat (perhaps a clean plastic container, preferably white).

2. Add 2 ounces of pearlash (potassium carbonate; washing soda may be substituted, but use only 1 ounce).

3. Place 2 ounces of natural indigo and one ounce of madder in a very fine weave little bag and put this into the vat. Place the vat in a warm place.

4. Rub the bag at least twice a day.

5. With luck, this vat will probably come into order somewhat more rapidly than the straight sig vat (No. 1).

6. Dyeing may proceed when the vat is green or yellow-green in color using the same methods as with the Sig Vat No. 1. Add more urine as the volume decreases.

7. When the dye becomes weak, add more indigo and madder.

Trouble-shooting the vat:

Follow the same directions as for other fermentation vats. Add small amounts of alkali if fermentation is too vigorous, the vat smells sour, and the color changes from yellow-green or green to bluish green or blue. If fermentation is weak or stops, add more madder, molasses, dates, sugar, or bran, but in small quantities.

Sig (Urine) Vat No. 3: Artificial Sig Vat

If one chooses to use this type of fermentation vat but does not want to mess with old urine, there is always the artificial sig vat. This may be set up using clear household ammonia, or urea, or a combination of urea and ammonia for the alkali; natural bacteria or yeasts in the air or commercial yeast for the fermentation; and a source of food for the microorganisms such as madder, bran, sugar, dates, or molasses.

I have had reasonable success with such vats: the odor is not nearly so offensive, and if a combination of ammonia and urea is used, the vat is sometimes as easy to maintain as the natural urine vat. If urea is used, it may be obtained from biological or chemical supply companies, or from a pharmacy.

The vat described is a modified version of that of Gerber (34), and the quantities are for an experimental one-gallon jug.

1. Powder 1 to 3 level teaspoons of natural indigo and place this directly into 1/3 cup of clear household ammonia, or place the indigo in a small, fine mesh bag and place the bag into the ammonia. The ammonia is the source of alkali, but it also does a good job of wetting the indigo. Allow the indigo to soak until fermentation in the main vat is active.

2. Fill the gallon jug 2/3 to 3/4 full of warm water and add 3 table-spoons of sugar or 5 to 6 tablespoons of good molasses or Karo syrup. Then add 1 package or cake of commercial yeast and stir well. Several pieces of madder root may be substituted for the yeast.

3. When fermentation is active, add the indigo and ammonia. If the indigo is in a bag, rub the bag against the side of the glass several times the first day to facilitate release of the fine indigo.

4. If the vat balance is correct, the vat color will change from blue to yellow-green in 24 to 48 hours, at which time dyeing may commence. If the indigo is confined in a bag the vat may have little or no sediment.

5. The dyeing procedure is exactly the same as for other indigo vats.

Trouble-shooting the vat:

• If fermentation is active and the vat color does not change from blue to yellow-green in 3 to 4 days, cautiously add additional ammonia, say 1 to 2 ounces, and stir. This will decrease the yeast growth but also provide the alkali necessary to dissolve the reduced indigo and neutralize the carbonic acid produced by the yeast.

• If the initial fermentation decreases markedly upon addition of the indigo-ammonia mixture, it may pick up in a day or two as the alkalinity from the ammonia decreases. On the other hand, if fermentation slows and stops after addition of the indigo-ammonia and does not start in 2 to 3 days, it may never do so, presumably because of some chemical poisonous to the yeast in the indigo. Before assuming this, however, add a little more nutrient, such as molasses, and reseed with more yeast.

Maintenance of the vat:

• This vat may be kept in order for some time by judicious addition of both ammonia or urea and a carbohydrate source. Let the color of the vat, level of fermentation, and odor be your guides. For dyeing, the vat should be yellow-green or green. Should the vat turn bluish green and fermentation be active (bubbles coming to the surface), and should it not smell like weak ammonia but rather sour, then ammonia or urea is needed. If the vat turns bluish green or blue and smells of ammonia, and if fermentation appears to have stopped, the vat contains too much ammonia and/or is deficient in nutrients. In this case, add more sugar or molasses, stir well, and permit the vat to rest. The sugar will accelerate yeast growth if that is the deficiency. If the alkalinity is too great for yeast growth, one or more days may be necessary until some of the ammonia is partially neutralized to ammonium carbonate from absorption of carbon dioxide by the vat. Yeast cannot grow in a solution that is too highly alkaline.

• Fermentation vats may be made temporarily inactive and kept, to save the indigo. Possibly the best way is to permit fermentation to continue until all of the carbohydrate source is consumed by the bacteria. Thus, fermentation will stop while the vat is still alkaline. A vat should not be permitted to become inactive by allowing it to ferment until sour, since once putrid fermentation takes over, the indigo may be irretrievably lost. If this appears to be occurring, add alkali of any source at once (ammonia, washing soda, potassium carbonate, or lye). It is preferable to keep the alkalinity high enough to prevent the vat from becoming sour but not so high as to prevent fermentation until the carbohydrate source is spent. The vat may then be stored (as long as it does not freeze) until rejuvenated. For rejuvenation, add alkali, carbohydrate source, and if needed, more indigo.

Fermentation Vat No. 4: Saxon Vat

In 1977 Dale Liles reasoned that sufficient nutrients, urea, microorganisms, etc., might exist in a dirty (unscoured) woolen fleece to permit reduction and solution of indigo by itself. Fred Gerber was consulted and he, too, suspected the method would prove feasible.

Indeed, the method worked quite well, the resulting dyed fleece being a very permanent and lovely blue. In 1980 I discovered (in Matthews, 418) that this method was one of the earliest used in Europe and that, at least up until 1918, it was still practiced by the peasants of Saxony and was described as the "celebrated Saxon blue." Note that indigo sulfonate (an acid dye) is/was also called Saxon blue, but is an entirely different dye. (According to Bancroft, vol. 1, 168, Counsellor Barth at Grossenhayn in Saxony was the first to have dyed cloth with indigo, after reacting it with sulfuric acid.)

1. Fill a plastic pail, say 5 to 6 gallons, with raw, unscoured fleece.

2. Cover with warm water (80° to 140° F). Soak for 24 hours. Remove and scour the fleece. Rinse well.

3. To the dirty water, add 2 to 3 teaspoons of well-ground natural indigo (or place the material in a fine mesh bag). Add the scoured wool.

4. Put a cover on the pail and set it in the sun.

5. Stir gently once each day and rub the bag if the indigo is so enclosed.

6. Fermentation should start in 2 to 4 days (bubbles rising to the surface). If it does not do so, add 1 or 2 dates.

7. In 7 to 10 days remove a lock of the fleece and air it for a few minutes. If it is dyed blue, wash it to see if the color is permanent.

8. If the color is not permanent, add 1 level teaspoon of urea and one or two dates and stir gently. Test another lock in a few days.

9. If the color of the first lock is permanent and the shade is dark enough, squeeze, remove, and air the wool on clean grass. Air the wool for at least 2 hours before washing, or possibly finishing off with a sour rinse and then washing.

10. If the color of the first lock is permanent but of a shade not dark enough to suit, place the wool back in the vat for a few more days. Then check the color of a lock again.

11. If the color of the first lock is permanent and of correct shade, and if an additional new load of scoured fleece is to be dyed, add 1 teaspoon of urea and 1 or 2 dates, stir, and add the new fleece. Additional indigo may or may not be necessary. Then proceed as before.

• As long as the vat ferments slowly and the indigo reduces and dissolves, no additions are necessary. Under these conditions the color of the vat water should eventually change from blue to green. If fermentation seems to stop, add 1 or 2 dates, which will release sugar slowly and also provide vitamins and minerals for the microorganisms. If fermentation occurs but the vat remains blue and the wool does not dye a permanent blue, urea must be added in small quantity.

• This vat is of particular value to wool spinners and felters since the shades produced are most stunning, and quite wash- and rubfast.

- This vat does not use lye and thus has appeal to those wishing to avoid its use.
- This vat, though restricted to wool fleece, may well be the easiest and most convenient of all the fermentation vats to set up and maintain. I recommend it most highly.
- Sometimes fermentation in this vat or any other type of fermentation vat does not occur. Suspected causes include chemicals in the indigo and sheep dip or fumigation chemicals.

Fermentation Vat No. 5: "Appalachian Vat"

The Appalachian vat is representative of vats using drip or wood ash lye (potassium hydroxide and potassium carbonate) for the alkali rather than ammonium hydroxide and carbonate from old urine. This was also known as the "potash vat" in earlier times. Vats using soda ash were called "soda vats" or "German Vats"; others used lime or mixtures of lime and pearlash or soda ash. The older literature abounds with these recipes, the most accessible being in Bemis and Bronson. Matthews lists a number as well.

The recipe given is virtually taken from Goodrich (13-14), though it is representative of many from the eighteenth and nineteenth centuries. It has worked well for me, coming into order in about 5 days, even at the low temperature of 75° F.

1. Use a half-bushel pot full of warm water (4 to 6 gallons). The typical pot in the Southern Appalachian Mountains was made of iron. A plastic wastebasket may be substituted. You will also need 1 to 2 ounces of natural indigo in a little tight-weave sack, 2 ounces of madder, 1 teacup of wheat bran which may be obtained at many supermarkets or at a health food store), and a half-pint of drip lye (see recipe below) — or enough to make the bran feel slick. The water will feel slippery or slick at about pH 9–10. I have used pearlash or soda ash instead of drip lye; two or three teaspoons of either one should be about right.

2. Add the warm water, bran, drip lye or washing soda, and about 1/2 ounce of the madder to the vat.

3. Drop or suspend the bag of indigo in the vat.

4. Keep the pot warm, in the sun, or on the hearth, next to the fire. Cover with a loose lid or cloth.

5. Rub the bag of indigo once a day to release the fine indigo and add more of the madder in 2 to 3 days. (The slow release of sugar from the madder and bran serves as energy source for the ferments.)

6. Wait until the dye "comes"; it may take a few days or up to 2 weeks. The sign of its "coming" is its turning green with a blue foamy scum. Odor is another indicator — at first a rather sickly smell but later on more pungent.

7. The dyeing procedure is the same as for other indigo vats. Enter only well-scoured materials. Since this vat has a sediment, articles should be suspended in the vat.

Trouble-shooting the vat:

- If fermentation is heavy but the vat does not turn green, add, cautiously, more lye.
- If the vat does not ferment, it may be because the vat is too alkaline. In this case the bran and liquid may feel quite slick. Wait a few days for this to decrease. If fermentation still does not start, add a little syrup. If fermentation still does not start, there may be something associated with the indigo which is poisonous to the microorganisms.
- Enter and remove materials from the vat very carefully. Addition of much oxygen to any fermentation vat will quickly render it in a nonreducing condition.

Maintenance of the vat:

- After each dyeing session, rub more indigo out of the bag and stir gently. The vat should be ready again in several hours. If much dyeing is done, additions of madder, bran, lye, and indigo will be required as time passes. Experience and proper judgment are necessary.
- Cotton or linen items are best dyed when the vat is strongest, and before any wool is dyed.
- Dates or raisins may be used as a sugar source in place of dried madder roots, but the microorganisms associated with madder are excellent ferments in starting a vat. Therefore, it is suggested that some dried woad or madder root be added when setting up the vat, at least.

Drip or wood-ash lye:

This is from Seymour (186).

1. Drill a number of holes in the bottom of a wooden barrel (if a plastic container is substituted, use one with a strong bottom and drill small holes).

2. Place a layer of gravel in the bottom, say 2 inches or more thick, and cover this with 3 to 4 inches of straw.

3. Fill the container with hardwood ashes (over the straw).

4. Pour rainwater, deionized, or distilled water (soft water) on top of the ashes.

5. Eventually liquid will begin to drip (slowly) through the holes in the bottom of the container. Collect this. The lye may be used as is or concentrated to a known specific gravity.

6. If the lye is to be concentrated, boil it down until it will just float a fresh egg. At this point, it will be quite strong (do not touch) and should be stored in a plastic bottle with a good lid.

7. Lye so concentrated may be used (add only small amounts at a time) in fermentation vats.

Fermentation Vat No. 6: Woad Vat

In modern times it is doubtful that many will be interested in additional types of fermentation vats. However, the woad (*Isatis tinctoria* L.) vat should be mentioned due to its historical importance.

Woad is easy to grow in northern climates and can easily become a weed pest. When Julius Caesar and his Roman legions invaded Britain ca. 55 B.C. they were greeted by Picts, an ancient Celtic group whose bodies had been painted blue with the indigo from woad. Apparently, this habit had been in practice for many years as part of religious practices and to make the warriors appear more fierce. A good discussion of this topic is in Brunello (108–109).

Indigo was dyed in Europe and probably Egypt with straight processed woad and alkali. The indican content of woad is very low compared to species of *Indigofera*, and it was little used after Indian indigo was introduced and accepted in Europe. Apparently, the Venetians were the first to use indigo, which was introduced in Europe in the 1100s from Baghdad (Brunello, 144). Later, in other parts of Europe there were laws and embargoes forbidding the use of indigo from India. This was, primarily, protection for the European woad growers. In fact, indigo was called the "Devil's Dye" by the woad growers, who also labeled it as a corrosive substance. There was some actual truth in this, in that indigo was reduced with orpiment (arsenic trisulfide) by some, following an Indian method in use at that time (ca. 1600–1700), but indigo could be reduced in a fermentation vat as well. Woad was still used quite late in fermentation vats, though not so much for its indigo content as for its nutritive content and associated "ferments."

Woad seeds may be obtained from some seed suppliers. It is easily grown and may be harvested, dried, and stored. A typical vat containing woad as one of the ingredients is as follows (essentially the method described by Matthews, 418–19):

1. The ingredients are for a 6- to 8-gallon vat.

2. Break up and soak 1 to 2 ounces of dried woad in enough water to cover for several hours.

3. Add the woad and water to 6 to 8 gallons of water heated to about 160° F.

4. Add 1/2 to 1 ounce of bran, 1 to 1 1/2 ounces of washing soda, 1 to 2 teaspoons lime, 1 ounce of dried madder roots, and 1/2 to 1 ounce of well-ground or pasted natural indigo (you might substitute 2 to 3 teaspoons of synthetic indigo paste, but it may not ferment).

5. Stir well, cover the vessel, and place it in a warm place for 24 to 48 hours. If all goes well, fermentation will become quite active, and the liquid should change from blue to yellow-green with a blue froth on the surface.

6. At this point, stir the vat well, and if fermentation is very active, add a little lime, say 1 teaspoon.

7. When the vat liquor is clear and yellow-green or green underneath the blue scum, it is ready for dyeing.

8. Skim off the scum (save this) and commence dyeing as for other indigo vats. Try to maintain the vat temperature at 110° to 120° F by keeping it in the sun (or next to the hearth).

9. At the end of the dyeing session, put the scum back in the vat and stir gently. The vat will probably be back in order the next day.

10. If no woad is available, try substituting 1/2 ounce of sugar syrup, but then make sure to add the madder, primarily to supply the necessary "ferments."

Direct Dyeing Indigo

Indigo dyeing where pattern was to be produced was done by tie, stitch, fold, or by the application of a resist substance such as paste or wax. These methods produced areas where the dye could reach and areas that it could not penetrate. The process was most tedious if only a small amount of blue design was desired and the rest of the piece had to be covered with wax. Of course, the wax had to be removed following the indigo dyeing. Eventually, two methods were worked out to dye the indigo directly, and these were the most difficult forms of indigo dyeing.

The initial method was done first to very limited extent in India; in Europe, during the eighteenth and nineteenth centuries, considerable time, research, and money were spent on developing methods for the process of painting and printing indigo directly. According to Pettit (123) the first method, called "pencil" or "brush blue," also went under the names of *bleu d'Angleterre, Englishblau, bleu au pinceau, Pinselblau, Schilderblau, topical blue,* and *bleu d'Applicacion.* Obviously, England, France, and Germany were all involved in its production.

Little success was to be realized until adoption of an eighteenth-century Indian method using orpiment as a reducing agent for the indigo. Direct dyeing of indigo required a concentrated indigo solution in alkali, such as lime water or wood-ash lye, thickened with honey, starch, or a gum, such as gum arabic or gum senegal. The reducing agent also had to be strong. Such conditions could not be obtained in a typical indigo fermentation vat. A strong, non-living reducing agent was needed, and orpiment (arsenic trisulfide) filled the bill admirably, having no drawback but its extremely poisonous nature. In fact, a number of Indian and European indigo dyers lost their lives using the material.

Orpiment, which exists in natural deposits, was traded from very early times because of its unique bright yellow color and thus its use (originally) as a pigment. Other reducing agents available reasonably early, but perhaps somewhat later, which could do the job were green copperas (ferrous sulfate), and tin oxides or stannous chloride. Later on, zinc dust could have worked, and, later yet, sodium hydrosulfite.

Whether the pre–eighteenth century Hindus possessed orpiment, oxides of tin, or copperas is speculative. In 1794, Bancroft suggested the use of glucose (honey) alone, and if reduction did not adequately occur, the addition of oxides of tin. I have found glucose alone to be entirely too weak a reducing agent.

By about 1730 dyers in England used a mixture of indigo, lime, potash, gum senegal or gum arabic, and a great deal of orpiment. The large amount of orpiment retarded the oxidation of the indigo long enough to permit application of the material rather successfully by penciling or with a small brush. Thus the name "pencil blue." The method was adopted rather soon thereafter in France and Germany (Pettit, 123–24, and Floud). A formula, originally from Cooper, and listed by Pettit (124), is as follows:

1 pound indigo
1 pound potash
1 pound orpiment
2 pounds lime

Grind these, boil in 3-4 gallons of water. To one quart of this solution add one pound of gum senegal. When in order for penciling, it is of fine green with a beautiful scum at the top.

I have altered this recipe, experimentally, with reasonably good success by substituting either thiourea dioxide or sodium hydrosulfite for the extremely poisonous orpiment and gum arabic for the gum senegal. My modification for half a pint of concentrate is as follows:

Pencil or Brush Blue: Cotton or Linen

1. Dissolve 2 to 3 ounces of gum arabic in 4 ounces of hot water. Heat, if necessary, until all goes into solution.

2. Dissolve 1 teaspoon of lye in 4 ounces of water. Add 2 teaspoons of synthetic indigo paste and stir well. More or less indigo and gum arabic may be needed, depending upon the material.

3. Add 1 teaspoon of thiourea dioxide or 2 teaspoons of sodium hydrosulfite and stir gently.

4. Add the gum arabic solution to the dye-indigo-hydros solution and stir gently.

5. Put a lid on the jar and try to keep the concentrate at 110° to 140° F until the indigo is reduced (concentrate is yellow-brown).

6. Apply the concentrate, by brush, to previously scoured dry cotton cloth. Move the brush from the concentrate to the cloth quickly; otherwise the indigo will oxidize to the blue stage on the brush.

• The high alkalinity of the concentrate will not harm cotton or linen, but it may destroy the luster in silk. If silk is used, make certain that it is treated with a sour afterbath.

• For cotton or linen alone, copperas or zinc dust and lime-water concentrates may be substituted for the concentrate listed above, but a thickener (boiled starch, gum arabic, or sodium alginate) should be added.

• The original recipe (Cooper) apparently did not work well for block printing. My modification may, since I believe sodium hydrosulfite or thiourea dioxide to be even more powerful reducing agents than orpiment.

• Apparently pencil blue was used quite a lot in the eighteenth and early nineteenth centuries, especially where small amounts of blue and green (here the yellow was printed first and the blue brushed over) were wanted along with other mordant-dyed colors in chintz patterns.

China Blue (English Blue, *Bleu de Faience*)

Pencil blue was the first method of painting or printing indigo directly on fabric, but it was always a difficult and somewhat imperfect procedure. Around 1785, after much thought and experimentation, another method was devised. The method worked so well that its use was quickly adopted in much of Europe. This method also permitted printing indigo directly with wooden blocks, copper plates, and engraved rollers. Thus, unless simple wooden blocks were used, a fair amount of equipment was necessary. The problem with pencil blue was that the indigo often oxidized on the brush or block before it was applied to the cloth, and any oxidized indigo would not dye cloth. The China blue method did not involve applying reduced and dissolved indigo. Instead, the method consisted of applying a mixture of very finely ground indigo and copperas, thickened with gum senegal. After drying for 4 to 5 days, the printed cloth was immersed several times, alternately, in a lime water vat and in a copperas vat. Thus, the indigo was applied in the blue state. Alternate immersion in the alkaline vat and in the copperas vat effected reduction and solution of the indigo in place.

Pettit (130) describes the method given in *Rees Encyclopedia* and by Cooper. I have experimented with the method a little and believe that it has distinct possibilities. My work has been with brush alone or using a brush through a stencil. I have not tried the method using a block.

The original recipe as copied by Pettit is as follows:

1. Indigo and copperas are ground with a small amount of water to the consistency of oil and thickened with gum senegal (1 part by weight of indigo to 1 1/2 parts good green copperas).

2. The thickened material is then printed on cloth.

3. Lime water and copperas solution vats are then prepared, and the cloth, after drying 4 to 5 days, is dipped, alternately, as follows:

A. Lime water—5 minutes
B. Copperas vat—30 minutes
C. Lime water—10 minutes
D. Copperas vat—30 minutes
E. Lime water—20 minutes

F. Copperas vat—45 minutes
G. Lime water vat—45 minutes

Following these treatments the dyed cloth is placed in a sour bath for 15 minutes to an hour before the final washing with soap and rinsing. Note: Cooper's directions, according to Pettit (131) are 1 part indigo, 1 part copperas, and 1 part gum water for dark blue, and 1 part indigo, 1 part copperas, and 5 parts gum water for a second lighter blue, apparently on the same fabric. In this case, the darkest blue was printed and allowed to dry (until the next day), and then the second blue was printed. Then the whole was allowed to dry for 4 to 5 days before immersion in the lime water and copperas vats.

My experiments with the method suggest the following recipes as starting points for experimentation:

China Blue Recipe No. 1: Cotton or Linen

1. Grind 1 teaspoon (2 grams) of synthetic indigo powder with 3/4 teaspoon (3 grams) of good-quality green copperas. Grind the copperas in the mortar dry first, and then add the indigo and a small amount of water and grind some more, until the mixture has the consistency of oil.

2. Dissolve 2 to 6 ounces of gum arabic (depending upon thickness desired) in 6 to 8 ounces of hot water. When the gum arabic is all dissolved, filter it through a fine mesh cloth. (A boiled starch solution, strong sugar solution, or sodium alginate solution can probably be substituted for the gum arabic.)

3. Add the indigo-copperas mixture to the gum arabic and stir well.

4. Print by brush or block to prescoured dry cloth.

5. Allow the cloth to dry and age for 4 to 5 days.

6. Add 1 tablespoon lime per gallon of water and stir well. Permit the lime to settle and decant the clear solution. This is the lime water vat.

7. Dissolve 1 to 2 teaspoons of good copperas per gallon of water. This is the copperas vat.

8. Dip cloth as follows:

A. lime water—5 minutes
B. copperas—30 minutes
C. lime water—10 minutes
D. copperas—30 minutes
E. lime water—15 minutes

9. Air for about 5 minutes and place the material in a strong sour vat for at least 15 minutes. (The method fixes the indigo but also produces iron buff on the cloth. This may or may not be desirable. A very strong sour bath will remove the iron buff.) (See discharge method for iron buff listed with the iron buff recipe in the Orange chapter).

10. Wash with detergent and dry.

I have also attempted to modernize the recipe as follows:

China Blue Recipe No. 2: Cotton or Linen

1. Grind 1 part synthetic indigo with 1 part sodium hydrosulfite or 1 part thiourea dioxide (with a little water).

2. Add gum water thickener.

3. Print. Dry overnight.

4. Alternate in lye solution (1/2 teaspoon per gallon) and then in sodium hydrosulfite (1 teaspoon per gallon) or thiourea dioxide (1/2 teaspoon per gallon). Leave in each solution for 5 to 10 minutes. Place in each solution at least two times.

5. Air for 1 hour and finish off with a good wash with detergent. Rinse well.

Saxon Blue (Indigo Carmine, Indigo disulfonic acid, "Extract," "Chemic," Indigo Sulfate)

Saxon blue was discovered shortly after development of concentrated sulfuric acid, around 1740 (see my earlier discussion of the two types of Saxon blue). It is an acid dye, not vatted indigo, and it dyes wool and silk directly, very well and in a striking greenish blue color, quite different in shade from vatted indigo. It was used very little on cotton and linen, and it is not as lightfast as vatted indigo. In fact, articles that will be subjected to strong sunlight or reasonably strong light for extended periods should not be dyed with it. It also produces the most striking greens, particularly in combination with "old fustic."

Preparation of "sulfate of indigo" extract (for wool and silk):

Most of the old recipes are basically the same, the difference being only in the amount of chalk added, if any. I generally add a very small amount; the extract will work fine with no chalk. Too much chalk will ruin the product.

1. Into 1 pound of concentrated sulfuric acid (9 liquid ounces) stir in slowly and by degrees 3 ounces of best quality natural indigo or 1 1/2 ounces (8.5 level tablespoons) of synthetic indigo. Make sure that the indigo is very well ground before adding it to the acid (this applies, primarily, to natural indigo). This should be done in a strong heat-resistant glass vessel. Use a glass rod for stirring.

2. Stir the mixture several times during the next few hours and keep the mixture at about 100° F, if possible.

3. The next day add gradually, stirring in slowly, about 1 teaspoon of chalk (optional).

4. Stir again the next day. At this point the preparation is ready to use. Bottle the extract tightly and it will keep at least a year or two.

5. I prefer to keep the product in a rather wide-mouthed bottle with a good, tight sealing lid so that the material may be removed by the spoonful.

Saxon Blue: Wool or Silk

1. Add 2 to 3 teaspoons (14 to 21 ml) of "sulfate of indigo" extract to 4 to 6 gallons of water and stir.

2. Add the well-scoured and wetted-out wool.

3. Heat slowly up to the simmer, say over 30 to 45 minutes, then keep at the simmer for 10 to 15 minutes. If the color is dark enough, remove, cool, and rinse. If the color is not dark enough, add more extract and reenter the wool.

Note: Some old recipes call for premordanting the wool with alum — 2 to 3 ounces of alum per pound of wool — or adding the alum to the dyebath, or rinsing the dyed material in alum solution (1 ounce alum in 4 to 6 gallons of water per pound of wool). These treatments may make the product more light- and washfast. According to Hummel (318) silk may be dyed with the straight, diluted extract or with a little sulfuric acid added, or it may be premordanted with alum and dyed as follows:

1. Dissolve 4 ounces of alum in 4 to 6 gallons of water at 120° to 140° F.

2. Steep the silk for 12 or more hours in the mordant as it cools to room temperature.

3. Squeeze, but do not rinse, and place the material in diluted extract solution (same concentration as for wool, depending upon shade desired) with addition of about 1 1/2 ounces of alum.

4. Dye at 100° to 120° F for 30 to 60 minutes.

5. Rinse and dry *or,* since the material is mordanted, the shade may be altered. If shade alteration is desired this can be done in the following ways: For a reddish or violet shade, add cochineal liquor to the dyebath. For deep blue, add a solution of logwood—not more than 1 1/2 to 3 ounces of logwood per pound of silk.

Saxon Blue: Cotton or Linen

This method was not considered very satisfactory and was used only for overdyeing yellows to produce green. The preparation in this case, was called "chemic."

1. Add 1 ounce of indigo extract to about 5 quarts of water.

2. Add finely ground chalk, slowly and gradually, stirring until the acid is exactly neutralized (pH 7). (In the days before pH meters or pH paper, neutralization was judged by taste, but this should no longer be attempted.)

3. Once the solution is neutralized, let it settle for about 8 to 10 hours and then decant the clear from the sediment. (This preparation was used only for dyeing green on cotton after the material had been dyed yellow.)

4. Add the well wetted-out yellow-dyed material to the "chemic" and dye at 120° to 140° F. Work until the desired green results.

5. Rinse and dry.

6. As with all indigo extract dyes, avoid subjecting the dyed material to excessive light.

Growing and Processing Indigo

For those wishing to grow and process indigo or to direct dye with green plant material containing indigo (the most primitive method), accounts are given in Buchanan (1987a, p. 118), Gerber (1977, p. 27), Miller (23), and Stanfield (21). In fact, Miller sells the seeds of Japanese *Polygonum tinctorium,* or if you buy her book, it contains a coupon which may be redeemed for a package of seeds. Stanfield describes, in detail, the processing and dyeing methods used in Nigeria with species of *Lonchocarpus;* and the basic traditional method of processing *Indigofera tinctoria* is described in *Indigo in America* (11) and *Blue Traditions* (14).

The processing methods used in India, America, and other places growing *I. tinctoria* produce a more concentrated natural product (up to 70 percent indigotin content) than is the case with the African

Lonchocarpus, Japanese *Polygonum* and European *Isatis.* However, all methods involve harvesting the mature plant and breaking and composting it wet for some time. Microbial enzymes remove and metabolize the glucose, and the indoxyl radicals plus oxygen combine into indigo (indigotin). Such composted material may then be formed into balls, dried, stored, and used directly in the indigo vat. These balls provide indigo, microorganisms, and nutrients for the microorganisms, and it is possible to produce a fermentation vat by adding the balls alone to an alkaline solution. The indigotin content of a composted ball may be as low as 1 to 2 percent.

Historically, the basic concentration method involved cutting the indigo plants and fermenting them in large vats for several days. Eventually all of the indoxyl radicals from the indican would be in solution. Then the fluid, and of course some of the particulate matter, was drained from these vats into other vats below where "beaters" beat air with large wooden paddles into the fluid. Thus, the indoxyl radicals plus oxygen reacted forming indigotin (blue state) and water. After a few days the insoluble indigo blue settled out. This process was hastened with the addition of lime. The water was then decanted or evaporated, and the "cake" allowed to dry. Once dry, the cake was cut into pieces and sold. Very high quality indigo, so produced, would be 40 to 70 percent indigotin; poor grades would be only 20 to 30 percent.

I. tinctorium requires a hot climate and long growing season. It was grown commercially in Charleston, South Carolina, and to a limited extent in Louisiana. *I. suffruticosa* may be grown in somewhat more northern climates. Japanese *Polygonum,* and European *Isatis* (woad) may be grown in northern climates. In fact, as stated, woad grows so well that it can become a plant pest. Excellent directions for growing the various species of indican-bearing plants are given by Buchanan in her book, *A Weaver's Garden.*

Indigo: Concluding Remarks

Nomenclature: A degree of confusion exists in the literature concerning the dyeing principles of indigo (indigotin) and of aniline. The occasional implication that these two substances are the same is entirely incorrect.

From early times (at least the eleventh century A.D.) indigo (the dye) was referred to as *Nil* by the Indians, Persians, and Arabians. Several centuries later, the Portuguese term was *anil;* the Italian *anilina;* and the English *aniline* (Brunello, 123). In 1760, Jean Hellot (Fr.) placed indigo and quicklime in a dry retort, applied heat, and obtained an impure oily distillate which he named *aniline* after the Portuguese name of the parent compound, *Anil.* This "aniline," later to be produced by destructive distillation of wood (Runge, 1834) from nitrobenzene (Zinin, 1842) and finally from the destructive distillation of coal in the gas works, would, starting in 1856, revolutionize the dyeing industry with production of the first synthetic dyes (Brunello, 277).

In 1865 Adolph von Baeyer (Gr.) started researching the chemical makeup of indigotin, and in 1868 he published the correct chemical formula for it. In 1880 von Baeyer produced the first synthetic indigo (indigotin) from ortho-nitro-cinnamic acid, but the expense was too great for use in commercial production. Synthetic indigo was first put on the market by the Badische Dyeworks (Germany) in 1897, at a cost slightly less than the natural product. In the early 1900s synthetic indigo used naphthalene or aniline as starting point. Either may be converted to indigo through a series of steps (Matthews, 433–34). Indigo (indigotin) is diindoxyl, $C_{16}H_{10}N_2O_2$. Aniline is aminobenzene, $C_6H_5NH_2$.

To produce a synthetic indigo less expensive than the natural product, the Badische company spent 10 years in research at a cost of 18 million marks. This was only slightly less than the entire capital of the Badische company at that time. In the 1890s practically all of the natural indigo came from the Bengal region of India, north of Calcutta. During these years Bengal had anywhere from 250,000 to 1 million acres in indigo cultivation. The cheaper synthetic indigo created economic disaster in Bengal, and by 1920 very little natural indigo was being grown or used (Brunello, 293–94).

Lightfastness: Indigo is classed as good with respect to lightfastness. This is true only if the dyeing is properly done and finished. It does fade somewhat, and if light shades are to remain good over the years, the indigo must be well penetrated, not just a surface job. One of the biggest mistakes a beginning indigo dyer can make is to dye a relatively light shade on cotton or even wool in one or two fast dips, run the yarn off into a twisted skein, and then expose the skein to light for an extended period. If this is done, you may end up with several shades of blue upon opening up the skein. Should this occur, and should you not want multishades, the article may be redyed. Interestingly enough, often the lighter areas will take up more of the dye, evening out the yarn.

Methods to improve the lightfastness of indigo are discussed in the indigo chapters in both Hummel and Matthews. According to Matthews (415), indigo-dyed articles were sometimes given an after-bath composed of very weak copper sulfate and weak acetic acid. The copper sulfate may make the shade somewhat greener as well as improving lightfastness. Sometimes the dyed material was steamed for 30 minutes following dyeing; this may also improve lightfastness and produce a slightly more violet shade.

With wool, the material was often rinsed in a sour, and then given a milling with soap and fuller's earth to remove every trace of non-fixed indigo. In addition, the goods were sometimes boiled in a weak alum or chrome and tartaric acid solution to improve lightfastness and rubfastness (Hummel, 316). Hummel also states that boiling piece goods (unmordanted) in barwood, sanderswood, camwood, or brazilwood solution may make the indigo faster to light, particularly at the cut edges.

Yellowing: Occasionally, an indigo-dyed article may become somewhat yellowish in time, particularly if the item is not washed and is of light intensity. Should this occur, wash the item in detergent; the yellow will wash out. This yellowing probably occurs in items not adequately soaked and washed following dyeing. If this bleaching effect has occurred, the item may need redyeing. Always make certain to soak all dyed items for 12 to 24 hours, then wash with detergent and rinse well.

Use of modern vat dyes: The extremely washfast and lightfast modern anthraquinone vat dyes are prepared and used just like indigo (the original vat dye) except that they require a much more strongly alkaline vat. Therefore, anyone who learns to dye with vatted indigo may use modern vat dyes as well. Because of the high alkalinity of the vat, vat dyes are normally used only on the cellulosic fibers.

Yellow and red vat dyes are useful for producing permanent greens and purples when overdyed with indigo. If this is done, the yellow or red should be dyed first and then overdyed with indigo. Reversing the process does not work well because much of the indigo will be stripped by the highly alkaline red or yellow vat.

Overconfidence: Finally, it is particularly important for the novice indigo dyer to remain humble. If not, said dyer may be in for a little fall. For sometimes it may seem that indigo still has not given up all of its secrets, that there is still a bit of magic. Occasionally errors are even made commercially. Apparently, blue denim for fading blue jeans is dyed in one relatively long, strong dip. On occasion, jeans are sold that crock badly, indicating that a great deal of the indigo in the dyebath was unreduced, and that the denim was inadequately scoured afterwards.

5 Red Dyes

It has long been possible to obtain outstanding reds with natural dyes. The best of these dyestuffs were available and in use at least by 2000 B.C. Brunello lists at least forty different species of plants and insects from which red, reddish-orange, or reddish-brown dyes were extracted (329–94). However, when it comes to extreme light-fast and washfast qualities, madder reds were and are unparalleled. Medieval tapestries, British red coats, Persian rugs, Egyptian mummy cloths and Coptic fragments, early American quilts, and Indian calicoes and palempores provide the proof. It may well be that madder (alizarin) reds are as lightfast as the best of modern dyes.

With respect to fastness, the cochineal, kermes, and Indian lac reds are not much inferior to the madder reds. These dyes are all derived from closely related insects (family Coccidae, scale insects), and the dyeing principles of all are closely related compounds, chemically.

Madder reds were much used on cotton, linen, wool, and silk, while the insect reds proved suitable mostly for wool and silk, though cochineal was used by the nineteenth-century cotton calico printers. Ancient Peruvian textiles, scarlet Venetian cloths, Gobelin tapestries, Persian rugs, and scarlet red coats contain the insect red dyes. It is likely that the shroud of Christ was dyed with Kermes (Brunello, 107).

Fancy reds of a more fugitive nature were produced from the so-called redwoods. These include the families of brazilwood and barwood reds. Newly made, the dyes are quite pretty; but they fade, usually to a reddish brown. They dye all natural fibers, most satisfactorily in combination with the non-fading madder.

Safflower reds dye cotton, silk, and linen substantively and were used from antiquity, particularly in India. Safflower reds also fade, but less readily than the redwoods, and the color does fade true, simply becoming lighter in shade with time. And two species of *Thelesperma* (plant family Compositae) produce very satisfactory reds on wool, cotton, and basketry material. These are American Hopi Indian dyes.

Virtually no common weeds or plants produce satisfactory red dyes. Some dyers achieve reasonable success with pokeberries on wool, but lightfast qualities are wanting. Bloodroot can be used with reasonable success on basketry material (and was used by the Cherokee Indians, Eastern Band).

However, madder patches can and have been grown in temperate climates. Dale Liles has had one going on our property in Knoxville, Tennessee, for the past ten years. One can attempt to rear cochineal scale insects on prickly pear cactus, but this is probably impractical since it takes 60,000 female insects to make 1 pound of material. Safflower can also be grown, though I obtain the relatively small quantities that I use from a health food store. The redwoods must be obtained from imports, and, in fact, if much dyeing is to be done, it would be wiser to purchase the imported red dyestuffs in virtually all cases.

Madder

Madder is an extremely ancient dyestuff, extending back at least to 2000 B.C. A number of genera and species of the family Rubiaceae contain the main dyeing principle, alizarin. The principal species used are listed in Table 2 (from Brunello, Gerber, Gittinger).

Table 2. Origins of Madder

SPECIES	COUNTRY OR AREA
Rubia tinctorium, perigrina, cordifolia, (al-lizari)	Levant, Persia, Near East, India, East Indies
Morinda augustifolia, citrifolia, tinctoria (al, ail, saranguy, or chiranjie)	Western India and Indonesia
Oldenlandia umbellata (chay or chayaver)	Coromandel coast of India and Ceylon
Relbunium microfillium, nitidum, tetragonum, hypocarpium, and others	Mexico, Chile, Argentina, South America generally

All species listed in Table 2 plus some others were used early on. In the eighteenth century chay, and particularly that growing in the

Kistna delta of Western India, was considered the best, at least for cotton calicoes, palempores, and Indian Turkey red. Chay had a high alizarin content, especially that grown on Kistna delta soils, which are very rich in calcium. Possibly all species of alizarin-bearing plants grow better and produce more dye if grown on calcium-rich soils. If grown on calcium-poor soils, the plants produce better dyeing results with addition of calcium to the dyebath.

Madder (*R. tinctorium*) became, eventually, the most cultivated species and was grown in very limited areas in Italy for wool dyeing as early as 50 A.D., and in Belgium, Holland, France, and northern Spain by the eighth century (Brunello, 130). By mid nineteenth century *R. tinctorium* was cultivated in very large acreages in the Levant, France, and Holland. Levant madder was exported primarily from Smyrna and Cyprus. Madder was cultivated on a small scale in certain places in the United States in the nineteenth century.

The dyestuff exists in the reddish-orange roots of the plant, and maximum alizarin content (2 1/2 to 4 percent) develops at about 3 to 5 years of age (Napier, 294). Thus, roots are generally dug at three-year intervals. The roots at that age are up to 1/4 inch in diameter and may be several feet long. After being dug, the roots are cleaned, dried, and broken. Dried madder roots have a pleasantly aromatic, sweet, earthy smell.

The roots are then stored or sold immediately, depending upon variety. In former times, the roots were sometimes ground to a powder, prior to selling, but many buyers would not purchase ground madder because it was too easily adulterated with brick dust, red or yellow ochre, sand, clay, or sawdust (Napier, 296).

In the nineteenth century a commercial preparation of madder called *garancin* was sometimes sold and used. Garancin results from treating madder roots with concentrated sulfuric acid. The acid digests much of the woody material, but does not destroy the dyestuff (alizarin). Thus 100 pounds of roots could be concentrated to 30 to 40 pounds of garancin, which reduced bulk for shipping and dyeing.

By mid nineteenth century madder had been investigated chemically for more than fifty years; indeed, more than any other "dye drug." This work resulted in discovery of five different coloring matters in the roots as follows: alizarin, purpurin, pseudo-purpurin, xanthin, and chlorogenin (Matthews, 496). Prolonged dyeing at high temperature dulls the color of the important constituent (alizarin), mostly by means of the xanthin (yellow) and the chlorogenin (orange). The coloring matters exist in the form of glucosides which are separated by action of microorganisms into the dyestuffs and sugar (glucose). The important dyestuff was named "alizarin" from the Arabic name for the madder plant red dye which was *al-Lizari* (Brunello, 123).

Alizarin is a polychromic dye; that is, it produces different colors with different single mordants or mordant combinations. On cellulosics, pinks and reds are obtained with alum; iron-alum combinations produce lilac and browns; and iron oleates and tannates give purple-gray, gray-browns, and near blacks.

Graebe and Libermann (German), in the Bayer laboratory, established the chemical structure of alizarin in 1868. They also succeeded in synthesizing it later on in the same year. Their method was to start with dibromoanthraquinone, produced from anthracene (distilled from coal tar). Fusion of dibromoanthraquinone with alkali resulted in alizarin (1,2 dihydroxyanthraquinone).

As with indigo some thirty years later, this remarkable synthesis made dyeing far less expensive and more exact than was the case using the cultivated plant, but also created economic disaster for madder farmers, particularly in France. As Brunello points out, this was the first synthetic dyestuff exceptionally fast to light and washing, and the synthetic alizarin was identical chemically to the fast alizarin in the plant. The synthesis of alizarin also permitted subsequent production of a number of other good synthetic alizarin-derived dyes, such as alizarin orange, red s, brown, and blue-black (Brunello, 287–88).

Madder Reds: Procedures

As mentioned in *A Practical Treatise* (304–5), by judicious use of mordants, madder can produce a number of extremely permanent colors, of which red is principal. But it must be emphasized that the operations are rather long, tedious, and exacting, at least with cotton and linen. Wool is the natural fiber most easily dyed with madder or alizarin, but even here great care is necessary.

The following rules should be strictly adhered to:

• Cotton and linen, particularly, should be very well scoured prior to mordanting. Do not use bleached material.

• For reds alone, no iron should be present in the mordants or dyeing vessels. Do not use chipped porcelain vessels that may permit iron salts to enter the dyebath.

• For best results, alum-mordanted items should rest (age) for several days prior to dyeing.

- Alum-mordanted items should be well rinsed several times prior to dyeing (to prevent crocking). With cotton and linen this is best accomplished with "fixing solution" (see general mordanting procedures).
- Madder (alizarin) penetrates fibers slowly. This is particularly true of the cellulosic fibers. Therefore, as a rule, longer times in the dyebath are necessary, and mordanting and dyeing twice is sometimes necessary.
- Sometimes very old (several years) stocks of madder dye poorly. Maximum dyeing potential of some madder varieties exists in the fresh material, while others prove superior after some ageing—up to 1 or 2 years.
- Madder red, and especially Turkey red, dyeing of cotton and linen is probably as exacting and complicated as any form of dyeing, ancient or modern. Shortcutting the process leads only to mediocre or poor results. In spite of these problems, East Indian dyers were producing very good madder reds on cotton at very early dates.

The following recipes are for dyeing 1 pound of material.

Madder Red Recipe No. 1: Cotton or Linen

1. Scour well, rinse, leave wet—or dry—and mordant with one of the basic alum cotton mordants. Work the material in the mordant well for a few minutes, then sink the material, and allow it to remain in the mordant for 6 or more hours.

2. Remove, squeeze out, and dry.

3. Mordant with tannin at the rate of 1 to 1 1/2 ounces. The tannin should be well dissolved and the bath at 120° to 140° F at the start. Work the material for the first few minutes, then sink the goods and allow to remain in the tannin for 10 to 12 hours. Sometimes the mordanting is accomplished better by dividing the quantity of tannin in half and mordanting twice, particularly with piece goods. This helps to insure even mordanting.

4. Remove the material, squeeze out, rinse, and hang up to dry, or proceed to the next step.

5. Mordant again with the same batch of basic alum. (Mordanting may be tannin-alum-alum, also.)

6. Remove, squeeze out, and dry as before, time permitting.

7. Dye at the rate of 1 1/2 to 2 pounds of madder or with 10 to 12 grams of alizarin pasted with water, or with part madder and part alizarin. If using madder, add a minimum of 3 gallons of hot water per pound of madder and allow to soak 12 to 24 hours. Pour the

liquid into the dyeing vessel. For maximal extraction of the alizarin, grind the wet roots in a blender. Add 1 to 2 gallons of hot water to the madder and heat to 180° F. Put a lid on the vessel and in about an hour (or longer) add the liquid to that already in the dyeing vessel. Repeat this process once or twice again. By this time most of the alizarin should be extracted from the roots. Alizarin is only slightly soluble in cold water and not greatly soluble in hot water.

8. Rinse the mordanted goods thoroughly, preferably in "fixing solution" (see general instructions for cotton mordanting). If "fixing solution" is not available, rinse and soak several times in hot water. The material must be well wetted out and free of uncombined mordant.

9. Thoroughly dissolve 1 to 2 teaspoons (7 grams) of tannin and 1 to 2 teaspoons (4 to 8 grams) of chalk or 1 to 2 teaspoons (3 to 6 grams) of calcium acetate in the dyebath.

10. Add the squeezed-out wet goods to the room-temperature (70° to 100° F) dyebath, work well for several minutes, then allow the goods to remain for 20 to 30 minutes before applying any heat. At the end of this time, if all is well, the goods will be dyed orange to orange-red.

11. Heat the bath slowly up to 160° to 190° F over a period of about an hour, and continue dyeing for an additional hour. If madder is used, never permit the dyebath temperature to drop or to exceed 190° F or the color may be dulled. If straight alizarin is used, the dyebath temperature may exceed 190° F, but superior results are usually obtained with the lower temperatures.

12. Remove the goods from the dyebath. The color should be brick red or black-red, about the color of a ripe, sweet cherry.

13. Cool, rinse slightly, and place the material in 4 gallons of water containing about 1/2 teaspoon of detergent. Lyesoap is fine if the water is soft. Bring the bath to the simmer or boil and maintain that temperature for 15 minutes to one hour. Keep a lid partially over the vessel. This process should clear and brighten the goods (remove the dun). A more bluish shade of red may be obtained by adding 1/2 teaspoon of washing soda to the brightening bath.

14. Rinse well and dry.

• If the material turns out to be a pale color or unevenly dyed the chances are good that the material was not adequately mordanted, particularly with the alum. It is also possible that the dyebath did not contain enough dye. In any event, the material may be remordanted and redyed. In fact, mordanting and dyeing twice was often done in the old days as matter of course. If redyeing is decided upon, proceed as follows: Remordant with alum, but not with the tannin, unless at half strength, and redye and finish off as before. A full, even shade should be achieved this time. If the dyebath seemed strong at the end of the first dyeing, it may be kept and reused after mordanting again, but mold may start growing on the surface in a few days if it is not kept cool.

• The preceding recipe may be altered with good results using tannin at half strength or omitting it entirely, but only if aluminum acetate mordant is used. Mordant twice with the same batch of newly made aluminum acetate, allowing the material to dry thoroughly 4 to 7 days following each mordanting. Then rinse well, preferably with fixing solution, and proceed with the dyeing. A recipe essentially the same as this is found in Kuder (202).

• More initial preparation is necessary for madder than for alizarin, but with madder it is easier to keep the dyebath in good order. An alizarin or madder dyebath should be slightly acidic for red cellulose dyeing. Under these circumstances the dyebath is orange, orange-red, or red with alizarin, and orange-red to red with madder. A madder dyebath will nearly always remain acidic on its own, since madder contains "madder acids" (madderic and rubiacic, Napier, 299). If a hot alizarin dyebath starts turning bluish red or purple, add small amounts, carefully, of any acid, including vinegar, until the dyebath color reverts back to *red*. If this is not done, the dyed material will have too much of a bluish or purple cast.

• At present, I prefer using straight madder, but for practical purposes more frequently use a madder-alizarin combination. Using straight alizarin alone does not have any historical resonance and seems like using any other synthetic dye. Certainly, the alizarin may give a purer color, but the complexity of the madder red, in my opinion, is more interesting and unique. The madder-alizarin combination reduces the cost and work and also seems to provide the complex color of the madder. Theoretically, a pound of madder should contain 11.5 to 18.00 grams of alizarin. Available stocks probably contain close to 13.5 grams (3 percent). Therefore, if 1 pound of madder is dictated, one may use 1/2 pound of madder plus 6 to 7 grams of alizarin.

• Probably the best way of dyeing with madder is that employed by virtually all of the early dyers. This involved placing the madder, broken up fine, in a cotton bag, and placing this directly in the dyeing vessel along with water and the wet mordanted goods. In this way, the alizarin is taken up by the goods as it dissolves, getting around the problem of alizarin's low solubility. The disadvantage is that a large dyeing vessel is required. Another commonly used method was to put the madder directly in with the goods, but constant stirring sometimes is necessary to avoid possible spotting.

• Generally speaking, better results are obtained in madder or alizarin dyeing by adding calcium acetate or chalk to the dyebath at the rate of 1 to 2 percent of the weight of the madder used. Calcium, aluminum, and alizarin form the red color lake. Addition of tannin at about 1 percent of the weight of the madder enhances exhaustion of the dyebath and provides additional mordant (Hummel, 346). Too much calcium or tannin, on the other hand, is deleterious. Therefore, if the dyebath water is *very* hard, calcium acetate or chalk need not be added. In this case, add a small amount of vinegar or dilute acetic acid.

• As previously mentioned, alizarin is only slightly soluble in cold water. When the required amount (pasted with water) is put in the cold dyebath initially, only a small fraction dissolves. Goods added to the dyebath take up the dissolved portion slowly, and then more dissolves. Also, more dissolves as the dyebath heats up. Materials placed in the cold dyebath will spot on the bottom from the undissolved alizarin unless they are moved about frequently. In order to reduce this tendency, I frequently add only half of the alizarin initially, and then add the other half when the temperature of the bath reaches 160° to 180° F. By this time much of the original will have been taken up by the goods being dyed. The alizarin added later should first be pasted with hot water. Placing the alizarin paste in a very, very tight mesh cotton bag is helpful.

• Madder or alizarin dyeing always works best when you start out with a dyebath temperature of 75° to 85° F and work the material for several minutes. Heat should not be applied until the material has been in the cold dyebath for 30 minutes. Once heat is applied, the temperature should rise slowly and not be permitted to drop until the desired final temperature (usually 160° to 180° F) is reached. The dyebath temperature should then remain fairly constant until the dyeing is completed. A number of old recipes call for a rapid rise to the boil during the last 15 minutes of dyeing, but I feel this is unnecessary, and, in the case of madder, it may dull the color. Cotton and linen should be in the dyebath 2 1/2 to 3 hours, and wool 1 1/2 to 2 hours.

• Mordanted materials should be well rinsed prior to dyeing. Otherwise, the dyed goods may not be rubfast.

Madder or Alizarin Recipe No. 2 ("Short Turkey Red"): Cotton or Linen

This recipe is essentially that from Matthews (369). It is not as pretty as real Turkey red, but it is a very good madder red.

1. Scour well, leave wet, then mordant with 1 to 1 1/2 ounces of tannin for 12 to 24 hours (see Recipe No. 1).
2. For best results, squeeze out, hang, and dry. Otherwise, rinse once and proceed to step 3.
3. Mordant well with strong aluminum acetate or Basic Alum Mordant No. 2, working the cotton or linen thoroughly and evenly for a few minutes, then sink in the mordant for 6 to 24 hours.
4. Squeeze out, and allow the material to hang and dry for 2 to 7 days (aluminum acetate) or 1 day (Basic Alum Mordant No. 2).
5. Mordant a second time, using the same batch of alum mordant. In former times, aluminum acetate mordanted materials were hung and dried in a well-ventilated oven at 120° F for 24 hours.
6. Treat with fixing solution at 100° to 120° F for 30 minutes, then rinse well. If fixing solution is not available, rinse well several times in water at 100° to 120° F.
7. Prepare the dyebath and proceed as in Recipe No. 1.
8. Brighten by simmering or boiling for 1 hour in water or a weak soap bath.
9. Rinse and dry.

Madder Recipe No. 3 ("Common Red"): Cotton or Linen

This recipe is essentially that listed in *A Practical Treatise* (319) and is typical of nineteenth-century recipes employing a combination of madder or alizarin and brazilwood. The color produced is quite nice, and the brazilwood in combination with the madder works very well. The madder tends to protect the fugitive brazilwood to a certain degree. The brazilwood may be omitted from the recipe, but the color, at least initially, may not be as bright.

1. Scour well, rinse, and work the goods for a few minutes in strong aluminum acetate. Then permit the goods to remain in the mordant for 12 to 24 hours.
2. Wring out and, for best results, allow the goods to dry and hang for 4 to 7 days.
3. Mordant with tannin, at the rate of 1 ounce tannin. Start with

warm tannin (120° to 140° F). Allow the material 10 to 12 hours in the tannin.

4. Squeeze out and, for best results, hang the material to dry.

5. Rinse very thoroughly or, better yet, treat with fixing solution (see general instructions for mordanting).

6. Dye with 1 1/2 pounds of madder or 10 to 12 grams of alizarin and 2 ounces of brazilwood (see Recipe No. 1 for preparation of madder or alizarin and dyeing procedure). Shave the brazilwood fine and simmer for about 1 hour in at least 2 quarts of water. Soaking the brazilwood overnight prior to simmering helps to extract all of the dye. Add the brazilwood liquor to the prepared madder bath.

7. Brighten with weak detergent or 1 ounce soap, preferably lye-soap, dissolved in soft water. Add the goods when the detergent or soap is dissolved and boil for 15 minutes.

• If using any basic alum other than aluminum acetate, mordant with 1 ounce tannin first. In this case, mordant tannin-alum-alum, using the same batch of alum mordant both times. Proceed with the combination madder or alizarin and brazilwood dyebath.

• Brazilwood may be added to Recipes No. 1 and 2 as well.

Turkey Red

The very best, fastest, and most sought-after red on cotton and linen from about 1600 to 1930 was Turkey red. Indian dyers probably produced relatively fast madder reds by about 2000 B.C., and they undoubtedly developed the Turkey red process also, though this is usually credited to the Persian, Turkish, and Greek dyers along the area of the Mediterranean known as the Levant in about 1600 A.D. The process was much more complicated than that for simple madder red. It originally involved some thirteen to twenty tricky steps to be executed over a three- to four-month period. Ingredients used in various of the steps (from 1600 to 1880) included cattle, sheep, or camel dung; rancid olive oil, castor oil, sesame seed oil, palm oil, fish oil, or lard; soda ash, tannin, alum, chalk, madder, and sometimes blood. The brightening process, which produces the brilliant fiery shades, included boiling the dyed article, sometimes for several hours under pressure with soap solutions, and sometimes with tin salts. Even the best dye houses (usually set up for Turkey red dyeing alone) had reasonably frequent failures. Turkey red was expensive, but if well done it would last until the yarn or cloth was in tatters, and it did not fade or bleed out on surrounding white areas.

I think it is most probable that the Indians first developed the process because it was they who were the first (by several centuries) to produce good madder reds on cotton, and because Turkey red simply involves additional steps which enhance the dyeing of cotton red with madder. So intricate was the process that it took literally every country in Europe 150 years to steal away and master this "Oriental" technique. The French and Dutch were apparently first to obtain it in Europe in about 1750. The color has been variously described as a soft blue-red with inner glowing fire and with shades from somber to luminescent. By 1900 the process had been shortened to about seven days, but only with quite elaborate equipment and a slight sacrifice in fastness. Turkey red is the most complex of any dye known, ancient or modern, and its entire chemistry has never been totally confirmed, mostly because it was entirely replaced by the much cheaper fast vat and developed reds by 1910–40, at which point further study of its chemistry ceased. It went under the names *Turkey red, Levant, rouge turc, rouge des Indes, Adrianopolis,* and *Adrianople red.* Adrianople (modern Edirne), an ancient Turkish city, was a major center for its production.

With adjective natural dyes such as Turkey red, meticulous mordanting is required, and this is particularly so with the cellulosic fibers which do not combine chemically with the metal mordants with nearly the ease as do the protein fibers. This lack of mordant attraction prompted the ancients, particularly in India where cotton was the major fiber, to attempt, first, to "animalize" the fiber in the hope of obtaining mordanting comparable to that of wool. This mention of "animalization" appears in the very early dyebooks, and is referred to in *Ciba Review* by Shaeffer (no. 39, 1941). Indeed, the cellulosic fibers were pretreated with the liquid insides of slaughtered animals, with dung solutions, with milk, and with nut gall extractions (which were once thought to be animal products). Actually, the ancients were on the right track: pretreatment with animal liquid insides, milk, dung, and nut galls does make the fibers more receptive to combination with metal mordants. The animal products contain albumen, fats, and gelatinous substances, and the nut gall, tannins. These substances are attracted and held by cotton, and all combine with metal mordants. It was later found that a number of plants were good tannin sources. These included galls, sumac leaves, cutch, divi-divi, myrobalans, and oak leaves. Thus, in India, and to a lesser extent in the Middle East (Phoenicia), South America, and the East Indies, dyeing with adjective dyes generally involved premordanting first with tannin and then with the metal mordant. These methods produced dyes on cotton and linen which were virtually as fast as those on wool and silk.

Again, sometime later, probably in India, and probably by accident, experimentation with emulsified oils as "fixers" for metal mordants started. How or when emulsified oils were first used seems lost in antiquity, but sometime, probably in the 1500s, it was learned that, with madder red dyeing at least, more brilliant colors were achieved using oil in addition to dung, tannin, and alum. In addition, reds so produced seemed to become brighter with subsequent washing. This probably encouraged experimentation for clearing and brightening the product by boiling after dyeing, which is essential in producing the brilliant, fiery, Turkey red shades.

Incidentally, these same or other methods apparently never produced such vivid colors with madder on wool. Napoleon in 1803 offered a reward of 20,000 francs to anyone who could produce, with madder, the vivid Turkey red colors on wool which were available at that time on cotton (Brunello, 268). The reward was offered because of fear that Spanish-imported cochineal, used in production of scarlet-dyed wool, might be cut off because of the first revolt of the Mexicans against the Spanish government.

Examination and experimentation with at least fifteen Turkey red recipes from dyebooks written between 1750 and 1946 reveal a number of variations, but all are based on similar principles as elucidated by Horsfall and Lawrie (131):

• The process is usually done on unbleached cotton.
• Oxyfatty acids from emulsified or sulfated oils are fixed, presumably chemically, on the cellulose at one stage or another of the process.
• The mordant (alum) is always "basic" (less acidic than alum used for wool dyeing).
• The presence of calcium salts is also necessary and probably enters the color lake.
• Steaming (following dyeing), in the presence of the modified oil, is essential. It is during this brightening process that the shade changes from brick red to bright red.
• Operations done following or with the steaming result in a further brightening of the shade. This is due partly to greater development of the color lake and partly to further clearing of the goods, especially if madder is used.

Basically, the early (1600–1870) process was as follows:

1. The cotton yarn was scoured well for several hours in soda ash or pearlash (piece goods were not dyed at first, only yarn).

2. The yarn was then rinsed and dried.

3. Next the dried yarn was "tramped" and soaked in an emulsion of vegetable oil or lard and soda ash or pearlash. Often dung was added, at least with the first oiling.

4. Following oiling, the material was wrung out evenly, and frequently left damp overnight and then hung in the air for about 3 days. Bright sunlight during the day, followed by dampening by dew overnight, seemed beneficial.

5. The oiling was repeated, usually 6 or 7 times. During this processing some of the fatty acids from the oil apparently oxidize and polymerize, and become firmly combined with the cellulose.

6. Now several washes with weak soda or pearlash took place. Usually the material was soaked in the alkali, wrung out, and hung up to dry for a day. These treatments were repeated until the material was free of all unmodified oil. Finally, the material was rinsed and dried.

7. Next the yarn was soaked for several hours in a weak tannin solution or a tannin-alum solution and dried. If a tannin solution alone was used, this was followed by mordanting once or twice with "basic" alum. Often the material was left damp overnight following the alum mordanting and then dried slowly.

8. In some dye houses the entire series of steps was repeated at this point, starting with the oiling. Otherwise, the material was washed in a fixing solution, which for many years was a dung solution. The dung solution was replaced with chalk or sodium phosphate by about 1870–80 (the important ingredients in dung are the calcium and sodium phosphates, which give a solution of about pH 8).

9. The material was then dyed, without drying. Dyeing involved use of 2 to 3 pounds of madder or the equivalent number of grams of alizarin per pound of cotton. As a rule, 1 percent of tannin on the weight of madder was used, and several grams of chalk or calcium acetate were also added. The amount of chalk or calcium acetate added was dependent upon the degree of natural hardness of the water. Often, a small quantity of ox blood was added as well (which did provide albumin, if nothing else).

Early failure of European countries to produce satisfactory Turkey red may have been partly due to climatic conditions (lack of intense heat and sunlight to alter the oil) and to lack of understanding of the importance of calcium salts in the dyebath. In India, particularly in the Kistna Delta region, both intense sunlight and calcium-rich soils and water were plentiful. Such conditions also existed in the Levant. Imported Greek dyers did not produce good Turkey red in France until the discovery of the necessity of calcium salts in the dyebath (calcium forms a part of the color lake). Blood was not essential, but many dyers thought it helped, and it probably did because of the mordanting qualities of its albumen content.

Sometimes the dyeing was done twice, with half of the allotted madder used in each dyeing. Remordanting with tannin-alum or basic alum preceded the second dyeing.

10. Next the yarn was ready for the clearing and brightening processes. The oldest methods of clearing involved boiling for several hours in a lyesoap bath containing small amounts of soda ash or pearlash. Later, this was done in a closed vessel containing a small hole in the lid under pressure. The material was usually brightened twice, and after about 1750 a small amount of tin salt was often added with the soap in the second bath.

11. Finally, the material was well rinsed and dried, often in strong sunlight.

Thus, the price of red.

One of the best discussions of the history of Turkey red dyeing is that of Schaefer (1941, pp. 465–90). Excellent discussions of the older and newer processes of Turkey red dyeing are in Hummel (425–51) and Matthews (366–72). The older processes are described in detail in *A Practical Treatise* (303–19), and a French process is discussed in *Dicks Encyclopedia* (39).

I have been experimenting with Turkey red dye for the past ten years and only recently consider my product to be superior in all respects. Most of the currently available recipes are summaries, leaving out important details that can be learned only by experience. Therefore, the recipes given will be ones that I have modified so that they work for me. The recipes will be given, in detail, as with indigo. It is my hope that you will attempt Turkey red both for the gratification and beauty of the product and so that this process, so important historically, will not be lost.

I have used three different oil mordants, whose recipes are as follows:

Oil Mordant No. 1: Castor Oil Soap

This mordant is easy to prepare, with materials readily available. The idea for this mordant is from Matthews (371), but no directions are given in that volume concerning preparation of the soap.

1. Dissolve 2 1/2 to 3 level teaspoons (16.5 grams) of lye in about 2 ounces (60 ml) of water (be careful!).

2. Put 4 ounces of castor oil (such as sold in drugstores – also available from chemical supply companies) in a vessel such as an empty plastic margarine tub.

3. Pour the lye solution into the castor oil.

4. Stir with a wooden, plastic, or glass rod for 5 to 7 minutes. The yellow color of the oil will fade as the mixture develops a light creamy consistency. The product will become thick as the soap forms.

5. Set aside (open). The soap will be fairly well solidified in about an hour.

6. In 2 or 3 days the soap will be hard and nearly dry. It is ready for use or may be stored. (Castor oil soap is sodium ricinoleate.)

7. Cut up the soap and dissolve it in 2 gallons of deionized or soft water. If soft water is not available add Calgon water softener before dissolving the soap. Add Calgon until the water feels slightly slick. (Do not dissolve castor oil soap in hard water as it produces a terrible soap scum.)

8. This is the "oil," but at this point is a bit too alkaline. Add acid or vinegar cautiously until the pH is 8 to 9 (most will do this with pH paper).

Oil Mordant No. 2: Emulsion Process

This is the oldest type of oil mordant. It is the most time consuming of the oils to use, but produces the most washfast results. The recipe given is in *A Practical Treatise* (319). Actually, it is Cooper's copy of a method formerly used in Manchester, England.

1. Dissolve 1 ounce (28 grams) of pearlash in 5 quarts of water (soda ash works as well).

2. Add, with stirring, 1 ounce (29 ml) of fish oil or Gallipoli oil. Gallipoli oil was inferior grade, rancid olive oil (*huile tournante*). Use old olive oil, if available. The mordant may be used the next day.

Note: Add 5 ml of oleic acid if available. It is probably the most significant ingredient in rancid oil. My preference is to use 1 1/2 ounces of soda ash and oil in this recipe.

Oil Mordant No. 3: Turkey Red Oil

In about 1870–80 Turkey red oil appeared. It was, and is, produced by treating castor oil or olive oil with concentrated sulfuric acid (sulfated or sulfonated castor oil). The apparent advantage of its use is that some of the chemical reactions which occur with oil on the fiber, in the sun and air or in a stove, have already taken place. Therefore, theoretically at least, less mordanting and mordanting time is required than is the case with emulsion oil or castor oil soap. Use of Turkey red oil has at least three disadvantages, however: I have found only one chemical company that sells it (Sigma Chemical); it must be kept refrigerated; and it is rather expensive (about $23 a liter). Monopol oil, sometimes called Turkey red oil, produces a nasty, sticky effect, and I doubt that it is made with castor oil or olive oil. Add 200 ml (7 ounces) of 70% Turkey red oil to each gallon of (preferably) deionized water. (This is slightly less than that called for in former times, but seems to give better results.)

Turkey Red Recipe No. 1: Cotton or Linen

1. Scour the cotton yarn or piece goods (unbleached) very, very well.
2. Rinse well and squeeze.
3. Work the damp or dry material well in the chosen oil mordant. The oil should be worked thoroughly and evenly through the material. Some old recipes indicate a very short time period for this and others a long period (up to several hours). A short period is probably sufficient since the oil is not chemically bound at this point; rather it is simply absorbed.
4. Remove goods from the oil bath, squeeze thoroughly, and place in a container to remain damp overnight.
5. Hang up to dry and oxidize for at least 3 days. This is probably best done by hanging or clothespinning the material on a nonabsorbent white plastic clothesline in the sun, so that the material dries by the sun's heat during the day and becomes dampened by the dew at night. After 1870 special stoves were constructed for drying the oiled items. These were ventilated and run at 100° to 140° F. Twelve to 24 hours stoving was the usual time period.

• Dung (1 to 2 ounces per gallon) was often added to the first oil mixture (emulsion oil). This probably aids the mordanting process slightly, because of the albumen it contains. The dung treatment may precede the oiling. In this case, the goods should be well soaked and hung up to dry before starting the oiling.

6. Oiling should be repeated at least 2 more times with Turkey red oil, and 4 to 6 more times with emulsion oil or castor oil soap (use the same oil).

7. Next the oiled goods should be soaked for several hours, rinsed once or twice if oiled with Turkey red oil or castor oil soap, and hung up to dry. The first soak should be in deionized or soft water if castor oil soap is used. Several short soaks in weak washing soda solution (1 teaspoon washing soda per 4 gallons of water) followed by rinsing, squeezing, and drying should be used if emulsion oil is the mordant. A new soda solution should be made up each time. This is done to ensure removal of all unfixed oil. *Unless all unfixed oil is removed, the dyed article will feel greasy and the dye will never be entirely rubfast.*

8. Mordant with tannin at the rate of 1/2 to 1 ounce technical tannic acid (tannin) per pound of cotton. Thoroughly dissolve the tannin (15 to 30 grams, 4 to 8 teaspoons) in a gallon of hot water. Add this to 5 to 20 gallons of hot water (100° to 130° F) and add the material. "Tramp" (work) the material several minutes and then sink it and mordant for 6 to 12 hours. (Tannin treatment was always used with emulsion oil in the past, but probably never with Turkey red oil. My experience indicates that it is helpful in all cases, particularly if the oil mordanting is not as well accomplished as in the old days.)

9. Remove from the tannin, squeeze, and dry. At this point the goods should be a light fawn color.

10. Next, mordant with one of the alum mordants (see section on mordants). If aluminum acetate is used, make up a recipe using 1 1/2 pounds of alum or 1 pound of aluminum sulfate and mordant twice, allowing 4 to 7 days drying and curing time after each mordanting. Work the material in the mordant for several minutes, then sink, and allow the material to remain for 6 to 12 hours. Squeeze and hang up to dry or allow to remain damp overnight and then dry. Of course, the mordant will have to be diluted sufficiently to cover the material, though aluminum acetate was formerly used at 5–9°/ Tw. Basic Alum Mordant No. 2 is far less expensive than aluminum acetate, and with Turkey red, works very well. If it is used, make up a quantity requiring 12 to 16 ounces of aluminum sulfate or 16 ounces of potassium alum. Mordant twice, but 2 days' drying between mordanting sessions is sufficient (assuming, of course, that

the material is dry in this length of time). The material should now be slightly gray colored.

11. The dyebath should be prepared using 2 pounds of madder or an equivalent amount of alizarin. I prefer 1 pound of madder and 10 grams of alizarin (see Madder Red Recipe No. 1 for dyebath preparation). Make certain that the dyeing vessel contains no iron contamination. A copper, stainless, or undamaged porcelain vessel is best. Add 1 to 2 teaspoons of tannin (previously dissolved) and 2 teaspoons of calcium acetate or 1 to 2 teaspoons of chalk. Add about 1/4 ounce (6 to 10 ml) of Turkey red oil as well, if available. Stir the dyebath well.

12. Treat the goods with hot (110° to 140° F) fixing solution (see recipes for fixing solution, Alum Mordant No. 1). The fixing solution removes unfixed alum and oil. Work the materials very thoroughly for at least 30 minutes and then rinse well several times.

13. Introduce the goods into the cold dyebath (70° to 80° F). Work thoroughly for 30 minutes. At the end of this period the goods should be dyed orange or pinkish red. The dyebath at this point should be orange to red. If the goods are not dyed at the end of this 30-minute period, mordanting was not successful and should be repeated, starting at the beginning.

14. Start heating the dyebath with low heat, bringing the temperature up to 160° to 170° F in an hour to an hour and a half. Continue dyeing at this temperature or slightly higher for an additional hour. Do not permit the dyebath temperature to exceed 180° F. The dyebath will be red if it reaches 160° to 180° F. Be careful to add vinegar to the dyebath if it starts to turn purple. This generally does not occur unless straight alizarin is used.

15. Remove from the dyebath, cool, squeeze, and rinse. Rinse with the rubber gloves off. If the fingertips and fingernails become only slightly stained, the chances are that you have a good, nongreasy product. (If the fingernails are red and feel greasy, boiling the product in detergent may or may not remove the unfixed oil. Sometimes reoiling, followed by drying and rescouring in fixing solution solves the problem. Following this treatment, the article may require remordanting with alum and redyeing.) The article may have a good color or only a brick red. Often the color is of a ripe sweet black cherry.

If the yarn is poorly or lightly dyed, or the piece goods lightly or unevenly dyed, remordant once or twice with Turkey red oil, emulsion oil, or castor oil soap, and then with "basic" alum. Then treat with fixing solution and redye. This usually helps even out unevenly dyed goods, improves goods that are well dyed but too light, and is highly recommended.

16. The decision, at this point will have to be whether to reoil, remordant with "basic" alum, and redye, prior to brightening and clearing. Condition of the goods should be your guide. If the goods appear to be satisfactory, proceed with the clearing and brightening. This process depends, in part, upon available facilities. (If an autoclave is available, first hang the material up to dry.) When dry, the goods should be placed on a white cotton sheet in the autoclave and given steam at 15 pounds pressure for 1 to 2 hours. When removed, the material should be transformed into the slightly bluish, fiery, Turkey red. The second choice for this procedure is a pressure cooker. Set the cooker at 15 pounds, add the goods and water, and let it run for 1 to 2 hours. If a pressure cooker is not available, boil the material in a partially closed vessel for 1 to 2 hours. Add a little detergent and 1 teaspoon of washing soda if a bluer shade of Turkey red is desired. (By the early 1900s Turkey red was sometimes steamed under pressures as high as 65 pounds according to Matthews, 372.)

17. Now give the article a good sudsing in detergent, rinse well, and dry. In former times, the sudsing was done with palm oil soap or a good natural soap. For this method, soft water must be employed. For greatest brilliancy and a luminescent shade, use 1/2 ounce of soap or a little detergent, 1 teaspoon of washing soda, and 2 to 3 grams (1/2 teaspoon) of stannous chloride in about 5 to 6 gallons of water. Boil the material for 1 to 4 hours. This is best done in a partially closed vessel. Rinse very, very well following this treatment.

Turkey Red Recipe No. 2: Cotton or Linen

This method is somewhat less complicated than Recipe No. 1.

1. Scour material well, rinse, and dry.

2. Mordant with strong aluminum acetate or Basic Alum Mordant No. 2 or a mixture of 80% Alum Mordant No. 2 and 20% aluminum acetate. Dry thoroughly and allow at least 2 days' aging before proceeding to the next step. Rinse in deionized or soft water.

3. Oil twice and dry thoroughly, preferably in the sun, using castor oil soap (see section on oil mordants). Allow several days' drying between oiling sessions.

4. Soak and rinse in deionized or soft water.

5. Mordant with tannin at the rate of 1 ounce of tannin per pound of material. Dry.

6. Without rinsing, remordant twice with the same batch of chosen alum mordant. Dry thoroughly between treatments.

7. Treat the material with fixing solution (see recipes for "fixing" solution, Alum Mordant No. 1), rinse well, and leave wet (this should be done just prior to entering the material into the dyebath).

8. Prepare a dyebath using 2 pounds of madder or 1 pound madder and 7 grams of alizarin or 15 grams of alizarin per pound of cotton. Add 1 to 2 teaspoons of tannin and 2 teaspoons of calcium acetate or chalk. Stir well. Add 5 to 10 ml of Turkey red oil, if available, and stir again.

9. Dye and finish off as in Turkey Red Recipe No. 1. If the color is light, remordant once with oil, then with alum, wash with fixing solution, and redye.

Turkey red: additional comments:

• Emulsion oils were often used at concentrations somewhat greater than specified in Recipe No. 1. For example, Hummel's recipe for 1 pound of cotton would be approximately as follows: Add 90 ml of rancid oil and 9 grams of dung (optional) to 2 1/2 gallons of water. Then add enough of a concentrated solution (all that will dissolve) of washing soda to bring the whole to a specific gravity of 1.01, or until an emulsion results which will not break when left overnight.

• Addition of 5 to 10 ml of oleic acid to the oil and water mixture, before adding the soda, will be helpful since it is difficult to obtain rancid olive oil. Free oleic acid is probably the important emulsive ingredient in the rancid *huile tournante*. I did not obtain good results with the emulsion process until I started adding it. Oleic acid may be obtained from your druggist or a chemical supply company. Sesame seed and castor oil should work as well as olive oil.

• To date, I have not experimented with an oil emulsion made from lard, but it was used quite successfully in India, Sumatra, and other places. I assume that melted lard was added to hot water, and concentrated washing soda solution added until a stable emulsion resulted. This method would produce the least expensive oil mordant, assuming that the modern additive which prevents lard from becoming rancid at room temperature does not cause a problem.

• By approximately 1880 Turkey red piece goods dyeing was often done (in Europe) by a process developed by Steiner (Hummel, 427). Thus, by this time, cloth was usually dyed by Steiner's process or with the then new Turkey red oil. In Steiner's process the cloth was impregnated with pure hot oil in one operation, after which the goods received several passages through weak solutions of washing soda. This required elaborate padding machines, but produced a red of exceptional brilliancy and intensity, compared to the emulsion process.

• The following "Test for good Turkey Red" is listed in *A Practical Treatise on Dyeing and Calico Printing* (314): "A good Adrianople Red supports for 10 minutes the action of nitric acid at 18° of the areometer, without suffering any sensible change. By letting it remain longer in the acid, or by employing a stronger one, the cotton becomes more and more orange, and finally loses its color. The simple madder reds, exposed to the same test, disappear in less than three minutes." An areometer is a hydrometer for measuring specific gravity, but I do not know the conversion factor in present-day terms. However, good Turkey red will withstand 5 to 10 minutes in a nitric acid solution made by diluting concentrated nitric acid by half with water (remember to carefully pour the acid into the water). The nitric acid solution will dissolve a little of the color, and the color of the yarn or cloth will become somewhat more orange in shade, but the item should still remain dyed for at least 5 minutes.

• It is possible to brighten with tin. This process is tricky but may add greater brilliance to the dyed product. One method is to add a small amount of tin salt (stannous chloride) to the soap solution–soda ash clearing bath as described in Recipe No. 1. Experimentation on my own has resulted in the following method, though a similar facsimile was probably used in former times.

1. Pour, with great care, 15 ml of concentrated nitric acid (or its equivalent in more dilute acid) into 4 gallons of room temperature water.

2. Add, and dissolve with stirring, 1/2 level teaspoon (2 to 3 grams) of stannous chloride. When dissolved, the solution should be clear.

3. Soak and work the material in this solution for a few minutes.

4. Wash well and dry.

• CAUTION: *Do not call your product Turkey red until it is—and there will be no question in your mind when you have it.* No dye color in all of history has been so misrepresented as Turkey red. The deception dates back to Boettiger's discovery of the direct dyeing synthetic Congo red in 1884. At that time representatives of German dyehouses traveled through India proclaiming how inexpensive and easy it was to dye red with the new material. And so it was, but Congo red faded badly—to a rather ugly brown—at which time the item had to be redyed. Congo red was so cheap, compared to Turkey red, though, that many workmen lost their jobs, particularly in the Turkey red dyehouses in Manchester, England.

Actually, from 1870 to 1930, several somewhat fugitive commercial and home dyes were pawned off as Turkey red. For example, Ramsey reports the story of a woman from Georgia buying Turkey red powder at the store in Calhoun between 1873 and 1907 (20), a substance that was probably Congo red. And the deception continues to this day. My quilting friends purchase fabric dyed so-called Turkey red. In many cases the shades are not even nearly correct, and washfast qualities are wanting. Some modern cotton dyes are extremely wash- and lightfast, so there seems little excuse.

Madder or Alizarin Pinks: Cotton or Linen

I have little experience in trying to obtain madder or alizarin pinks on cotton. Hummel (451) recommends using weak aluminum acetate or straight aluminum sulfate (not basic) mordant with weaker oil mordant preparations and a weaker alizarin dyebath compared to Turkey red. He also recommends using a "blue shade" of alizarin.

I have produced 1 or 2 yards of a rather pleasing madder pink by premordanting with aluminum acetate which had been used several times for red. The dyebath was made up to near normal strength, and the dyeing procedure was exactly the same as for madder red.

Another good madder pink was produced by premordanting with tannin (1 ounce per pound of cotton), followed by copper sulfate (1/2 ounce), or weak Basic Alum Mordant, and then dyeing with madder (6 ounces).

Pinks my be gotten with safflower and cochineal as well, but neither will be as lightfast as the madder.

Silk Dyeing with Madder and/or Alizarin

Madder was not much used in silk dyeing because of its general lack of brilliance. However, a very pleasing deep red may be obtained using alum and madder or alizarin alone on some silks, and other pleasing reds may be obtained with madder-cochineal or madder–black oak bark combinations. (All recipes are for 1 pound of silk.)

Madder Red: Silk

1. Scour well, or wet out the silk in the case of scarves or piece goods.
2. Mordant for 6 or more hours with basic alum or 1 hour or longer with aluminum acetate or aluminum sulfate.
3. Dry carefully.
4. Remordant, unless new, strong, mordant was used.
5. When dry, rinse well and leave wet.
6. Prepare a dyebath with 8 to 12 ounces of madder or 8 to 10 grams of alizarin. Add 1 teaspoon of well-dissolved tannin and 1/2 to 1 teaspoon of calcium acetate or chalk. Stir well.
7. Add the silk and work well for several minutes at room temperature.
8. Apply heat and gradually increase the dyebath temperature to about 180° F over the course of 1 to 2 hours.
9. If using alizarin alone, add, cautiously, vinegar or other acid if the dyebath color changes from orange or red to a purplish shade.
10. (Optional) Remove, rinse, and place silk in a weak natural soap bath containing about 1 gram (1/4 teaspoon) of dissolved tin (stannous chloride). Heat up to nearly the boil for 10 to 15 minutes.
11. Remove, rinse, and dry.
12. If the results are unsatisfactory, redye with brazilwood—about 2 ounces. Soak the brazilwood chips overnight, then boil for one hour to extract the dye.

Madder–Black Oak Bark Scarlet: Silk

I have obtained very beautiful scarlet reds on silk with considerable luster using this combination. In fact, it is my favorite method of obtaining scarlet on silk. (It may not work on some silks, however. Try test samples first.)

The proportions and amounts of madder or alizarin and mordants are exactly the same as for madder red on silk (preceding recipe) except that the tannin is omitted from the dyebath, and a very small amount of black oak bark liquor added. Liquor from 1/8 ounce of bark may be enough for 1 pound of silk. Too much bark will produce orange. Therefore, add the bark liquor cautiously, and preferably after the goods have been in the dyebath for 30 minutes or so since the mordant picks up the yellow more rapidly than it does the red.

Madder–Cochineal Red: Silk

Madder-cochineal combinations produce my favorite reds on silk. The red produced varies depending upon the proportions of madder and cochineal used. The following recipe, a modification from Mairet (59), produces a nice, lustrous, bright red with a slight bluish cast.

1. Scour well, or wet out the silk in the case of scarves or piece goods.

2. Prepare a mordant bath by dissolving 1 3/4 ounces (5 level tablespoons) of oxalic acid in a minimum of 5 gallons of room-temperature water. The mordant-dyeing vessel should be nonreactive or brass or copper.

3. When the acid is dissolved, add and dissolve 1 3/4 ounces (3 level tablespoons) of tin (stannous chloride).

4. Add the wet silk and heat the bath up to 160° to 180° F. Mordant at this temperature for 45 to 60 minutes.

5. Next temporarily remove the silk, squeeze out, and set aside.

6. Pour away about half of the old mordant liquor and add the dye-liquor from 6 ounces of cochineal bugs and 6 ounces of madder roots or 5 grams of alizarin. Fill the container back up to 5 to 6 or more gallons with water if the cochineal and madder liquors are of less volume than the half of the mordant liquor that was discarded.

7. Reenter the silk and work well for several minutes since the dye will be taken up rather quickly.

8. Heat the dyebath back up to 160° to 180° F and dye at that temperature for about 30 to 45 minutes.

9. Remove the silk, squeeze, cool, wash well with detergent, rinse, and dry.

• The amount of cochineal called for in this recipe does seem a bit excessive. In fact, Mairet's recipe calls for even more (12 ounces cochineal per pound of silk). I have produced virtually the identical shade and depth on a second batch of silk by premordanting with the tin-oxalic acid, squeezing out, and dyeing in the same dyebath. I did add about half more of the originally suggested amount of madder liquor. It is entirely possible that good color may be obtained using only 4 ounces of cochineal, but I have not tried this. In any event, do not throw away the dyeliquor as it should dye wool well, even if it is too weak to dye silk to the desired depth.

• Since the color produced by this dyebath is a bluish red, it over-dyes nicely with indigo for a purple. If dissatisfied with the color, overdye with brazilwood, 2 ounces per pound of silk.

Wool Dyeing with Madder and/or Alizarin

It is possible to achieve bright to somber colors ranging from orange to bright red on wool, the best colors, in my opinion, resulting from use of madder or madder-alizarin combinations. If straight alizarin is used, addition of small quantities of a good yellow dye is often helpful.

Madder Red: Wool

  1. Scour wool well.
  2. Mordant with alum (3 ounces of alum and 1 ounce of cream of tartar).
  3. Allow mordanted material to remain damp for 1 to 3 days.
  4. Prepare a dyebath of 5 to 15 gallons, using madder at the rate of 8 ounces per pound of wool (or 4 ounces of madder + 4 grams of alizarin or 7 grams of alizarin + a small amount of yellow dye.) One level teaspoon of alizarin weighs about 1 gram. See Madder Red Recipe No. 1: Cotton or Linen for instructions concerning dyebath preparation.
  5. Unless the water is very hard, add 1 teaspoon chalk or calcium acetate. Stir well.
  6. Rinse the mordanted material well.
  7. Add the wool and work it for several minutes in the room-temperature to warm dyebath. One good way of having a warm dyebath is by heating the madder/alizarin to about 170° F in one vessel and then pouring this into the actual dyeing vessel, partially filled with cold water.
  8. Slowly heat the dyebath to not more than 190° F over a period of about 1 1/2 hours. If after 1 hour the color of the goods or dyebath seems weak, and the temperature has reached 160° F, additional pasted alizarin or madder may be added. If alizarin is added, to avoid spotting work the material until the alizarin appears to be dissolved.
  9. Remove the wool, cool and air, and rinse.
  10. If the wool is not red, add clear lime water to the dyebath. Approximately 8 ounces is usually about right when using madder or madder-alizarin combination. This frequently makes the dyebath redder. Often, this step must be omitted with a straight alizarin dyebath as it may change an already alkaline dyebath to bluish purple.

The dyebath should be red. Add a small amount of acid if the dye-bath color changes to bluish purple. The acid will change the color back to red; too much acid will change the color to orange. To make the lime water, mix 1 quart of water with 1/2 ounce lime (calcium oxide) and allow to settle. Use only the clear. If lime is not on hand, judiciously add small amounts of washing soda.

11. Reenter the goods; work well, and dye for 10 to 15 minutes longer at 170° to 190° F.

12. Remove, squeeze out or drain, and wash in detergent when cool, or brighten by heating the wool in a lyesoap afterbath. If this is done, use soft water to avoid soap scum.

• Twelve to 16 ounces of good quality madder will give lacquer red, 8 ounces will give bright red, 4 ounces, orange-red, and less than 4 ounces, orange on most wools. A few wools never seem to produce better than "tomato soup" red.

• Frequently the color is slightly bluish red, an altogether pleasing and very durable color.

• Very fine wools often require only 2 ounces of alum per pound of wool; woolen broadcloths often require 4 ounces of alum.

• Addition of 1 teaspoon of tannin frequently helps exhaust the dyebath.

• Addition of 1 teaspoon of tin to the hot dyebath will brighten the color, though often to a slightly orange-red. Dissolve the tin in about a quart of the hot dyebath liquor, briefly remove the wool, stir in the tin, and reenter the goods. Dye for an additional 20 to 30 minutes.

• Chrome-mordanted wool gives garnet red. Tannin and tin may be added. Tin-tartar mordant produces orange; copper-tartar, brown; and iron-tartar, darker brown.

Cochineal, Kermes, Lac, St. John's Blood (Polish Cochineal)

All of these dyes, like madder, are of extremely ancient use, probably dating back to approximately 1500 B.C. in Mexico, Central and South America, Egypt and the Mediterranean, and India. Cochineal was used as red food dye until about 1940, and for scarlet British army redcoats into the beginning of the twentieth century (Pellew, 92).

It is quite probable that the cochineal, logwood, old fustic, and brazilwood dyes were ultimately the most economically significant materials the sixteenth-century Spanish explorers brought from the New World. Eventually, even precious metals became of secondary importance.

The Dutch physicist Drebbel is generally credited with accidentally discovering in the early 1600s how to dye scarlet with cochineal, using tin salt mordants (Brunello, 201). Up to that time, or at least from the tenth century, alum-mordanted kermes was in general use in Europe for the red and scarlet Gothic tapestries, but by about 1650 the cochineal scarlet was in use. According to Mairet (31) the use of cochineal resulted in a brighter, cheaper,and uglier scarlet. At any rate, the earlier kermes-dyed tapestries have proved to be more lightfast.

The dyes are produced by closely related species of scale insects (family Coccidae) and are very similar chemically (anthraquinone derivatives). The cochineal insect possesses the highest dye content and is the easiest to use. Cochineal also produces the brightest colors, though they are slightly less lightfast than those produced with kermes or lac. All are considered to possess quite good lightfastness, though.

The several species of cochineal insects (*Dactylopius* sp.) feed on nopal or opuntia (prickly pear) cacti in Mexico, Central, and South America. The dyeing principle is carminic acid, $C_{22}H_{19}O_{13}$. Kermes scales (*Kermococcus vermilis*) feed on an oak shrub indigenous to the Mediterranean region. The dye is kermesic acid, $C_{18}H_{11}O_9$. Lac scales (*Lakshadia chinensis* and *communis*) produce the dye, laccaic acid, $C_{20}H_{12}O_9$, as well as shellac. Lac insects are indigenous to India, Tibet, China, Burma, and Siam. A number of species of plants serve as host for the lac insects. St. John's Blood was used for some time by medieval dyers (prior to the introduction of cochineal) in Poland, Lithuania, and Russia. This scale insect feeds underground, on the roots of what Bancroft called "German knot grass" (Gerber, 1978a, pp. 16–19). Apparently, the dyestuff is chemically similar to the others.

So far as I am aware, kermes and St. John's Blood are no longer available to the craft dyer, and it is usually difficult to obtain lac. Cochineal is expensive, but on wool, at least, it goes a long way. Like madder, cochineal is polychromic, producing many colors and shades that vary with different mordants, mordant combinations, and the pH of the dyebath. Generally speaking, cochineal produces brighter reds (scarlet) on wool and silk than does madder. There is less application for use of cochineal on the cellulosic fibers.

The best source of information on this group of insect dyes is *Cochineal and the Insect Dyes* by Gerber. Brunello also has a good discussion, and the best of the old literature is in Bancroft, vol. 2, and Hellot/Macquer. (All recipes are for 1 pound of material.)

Cochineal Pink: Cotton or Linen

1. Scour well and mordant twice with aluminum acetate or Basic Alum Mordant No. 2. Tannin-alum-alum mordanting produces crimson. Dry for 2 days or more between mordanting sessions.

2. Scour well with fixing solution and rinse well.

3. Prepare a room-temperature dyebath of 4 to 6 gallons of water and add the dyeliquor extracted from 2 to 4 ounces of well-ground cochineal.

4. Add the cotton, work well for several minutes and then slowly heat the dyebath to about 160° F. Keep at this temperature until the color develops.

5. Remove, squeeze, rinse well, and dry.

Note: The remaining dyebath may be used for dyeing wool or silk.

Cochineal Scarlet Recipe No. 1: Silk

I do not consider this recipe to produce as good a cochineal scarlet on silk as the madder–black oak bark combination. Neither is it as good a color as cochineal scarlet on wool. However, color preferences and needs differ markedly among individuals. The proportions listed are from a suggestion by Hummel. It is a one-pot method, and the recipe works well.

1. Extract the color from 2 to 4 ounces of well-ground cochineal insects. Soak overnight in 2 to 4 quarts of water, then heat to near boiling the next day for about 30 minutes.

2. Prepare the dyebath by adding 4 to 6 gallons of room-temperature water in a nonreactive vessel.

3. Dissolve 1 ounce (30 grams or 10 level teaspoons) of cream of tartar in the dyebath.

4. Add 13 grams (2 1/2 level teaspoons) of tin (stannous chloride). Stir until dissolved.

5. Add 13 grams (2 1/2 level teaspoons) of tin (stannic chloride). Stir until dissolved.

6. Add the cochineal liquor.

7. Add the scoured, well wetted-out silk. Work for a few minutes,

then heat the dyebath up slowly to 160° to 180° F. Keep at this temperature for about 1 hour.

8. Remove, squeeze, wash, rinse, and dry.

Cochineal Scarlet Recipe No. 2: Silk

In the nineteenth century, cochineal scarlet on silk was often done by first dyeing a deep annotto orange, then mordanting the material with tin spirits, and finally dyeing with cochineal. Essentially it was annotto orange overdyed with cochineal red.

1. Into 4 gallons of water put 2 ounces of annotto, 2 ounces of pearlash or soda ash, and 1/2 ounce of old-fashioned soap (lyesoap).

2. Apply heat and stir until all is dissolved. Bring the solution up to the boil for 30 to 60 minutes.

3. Let the temperature of the annotto solution drop to about 140° F.

4. Add the well wetted-out silk and work well in the solution for about 20 minutes.

5. Remove, squeeze out, and rinse in water twice.

6. Prepare a mordant solution by dissolving 1 3/4 ounces (8.5 level teaspoons) of tin (stannous chloride) in 4 to 6 gallons of warm water (100° to 120° F). Work the wet, annotto-dyed silk in the mordant for 1 hour.

7. Remove the silk, squeeze, and work the material in a 100° to 120° F dyebath (4 to 6 gallons) containing the dyeliquor from 4 ounces of cochineal. After working the material for several minutes, sink the material, and allow it to remain in the dyeliquor for several hours.

8. Remove, squeeze, rinse, and dry.

Cochineal Scarlet Recipe No. 3: Silk

This recipe is taken directly from Mairet (32).

1. Place 4 to 6 gallons of water in a nonreactive pot.

2. Dissolve 1/8 ounce (1 teaspoon) of oxalic acid in the dyebath. Add 1 ounce (5 teaspoons) tin (stannous chloride). Stir until dissolved.

3. Add the liquor from 4 ounces of cochineal. Stir well.

4. Add the well-scoured and wetted-out silk.

5. Heat up to the simmer and keep at that temperature for 1 hour.

6. Remove, squeeze, rinse, and dry. (If the color is weak, a very nice ruby can often be obtained by overdyeing the material at 160° to 180° F with the dyeliquor extracted from 2 ounces of brazilwood.)

Cochineal or Lac Scarlet, Wool

Cochineal and lac scarlets were produced both by premordanting and by one-pot methods. Good results were obtained both ways, with care. According to Hummel (351), the premordanting method gives a slightly bluer shade of red, requires less cochineal, and gives both purity and intensity of color. The one-pot methods yield a yellower and more brilliant scarlet. Most, if not all, of the seventeenth- and eighteenth-century recipes used the one-pot method, with tin "scarlet" spirit mordants added directly to the dyebath. Some recipes called for the addition of a small amount of yellow dye, usually turmeric, black oak bark, young fustic or Persian berries. According to Bancroft (vol. 1, 361), "at least 1/4 less cochineal is required if a small amount of black oak bark (quercitron) is added." The author of *A Practical Treatise on Dyeing and Calico Printing* devotes 9 pages (372–81) to discussing the production of scarlet and includes the methods of several experts of the time.

Cochineal Scarlet Recipe No. 1: Wool

This recipe, also called "fire scarlet," "Dutch scarlet," and "gobelin scarlet," is a modification of Hellot/Macquer (123–33) and Bronson (169–72). It is my favorite eighteenth-century method of producing scarlet, and requires little cochineal.

1. Scour the wool well, rinse, and leave damp.
2. Heat 5 gallons of deionized or soft water to about 180° F in a stainless steel, nonreactive, or block tin vessel. Do not use copper or brass unless it is scrupulously clean – and even then it is not the best since copper salts have an adverse effect on the color.
3. Add 1 ounce of cream of tartar and stir until dissolved.
4. Grind 1/3 to 1/2 ounce of cochineal well. Add this to the hot dyebath. Stir well. Optional: Add a small amount of yellow dye such as black oak bark, old fustic, or turmeric. Suggested quantities are 3 grams of turmeric, 2 grams of black oak bark, or 3 grams of old fustic. The yellow dye may be added initially, but preferably near the end of the second dyeing if deemed necessary.
5. Heat to the simmer, simmer 20 minutes, and add 1 ounce of scarlet spirits (Napier, no. 29; see Appendix B). Mix well.
6. Enter the wet yarn or piece goods. Move the material frequently. Dye at the simmer for 1 1/2 hours. The dyebath at this point should be virtually exhausted of color.

7. Remove the wool, drain, cool, and air the material well for about 15 minutes.

8. Rinse in cold water. The material should be dyed a fairly good orange scarlet at this point.

9. Empty the dyebath, and add 5 more gallons of soft water.

10. Mix 1/2 ounce of fine starch in warm water. Add this to the dyebath.

11. Bring the dyebath temperature to about 180° F again.

12. Add 1/3 to 1/2 ounce of ground cochineal, then add 1/3 to 1 ounce of scarlet spirits. Stir well.

13. Reenter the wool at the simmer. Dye for 1 1/2 hours.

14. Remove, cool, rinse well, and dry.

• If the cochineal is of good quality, only 2/3 ounce will be required for 1 pound of wool.

• If the cochineal has been allowed to become damp in storage, it may be fit only for purples, no longer being capable of producing good scarlet or reds at all.

• Add the scarlet spirits cautiously for the second dyeing, using the color of the dyebath as a guide. Addition of too much spirits may make the dyebath too orange.

• If scarlet spirits are not available, substitute 3/8 to 1/2 ounce of stannous chloride. Dissolve the tin in soft water. Add half to the first dyeing and half to the second dyeing.

• If the color is too flame orange, add a little alum, previously dissolved in hot water, to the dyebath near the end of the second dyeing. Another method is to dissolve 1 to 3 teaspoons of washing soda in 10 gallons of hot water. Work the material in this solution for a few minutes to 1 hour and then rinse very thoroughly. This should change the shade to more red and less orange, if done very carefully.

Cochineal Scarlet Recipe No. 2: Wool

This is my modern method of producing scarlet, and I believe it to be as good as any of the older methods. In addition, since it is a one-pot method, it is most economical in terms of time.

1. Scour well, rinse, and leave wool damp.

2. Heat 5 gallons of deionized or soft water to about 180° F in a nonreactive vessel.

3. Add 1 ounce of cream of tartar and stir until dissolved.

4. Grind 1 ounce of cochineal well. Add this to the hot dyebath. Stir well.

5. Maintain the dyebath temperature at 180° to 190° F and add 1/2 ounce of tin (stannous chloride). Stir until dissolved.

6. Enter the damp yarn or piece goods. Move the material frequently. Dye at the simmer for 1 1/2 hours. The dyebath at this point should be nearly exhausted of color.

7. Remove the wool, drain, cool, rinse well, and dry.

- If the color is correct, but somewhat weak, add 1/4 to 1/2 ounce additional ground cochineal to the dyebath, reenter the wool, and dye for an additional 30 minutes. Dye content of various batches of bugs differs somewhat. Too much cochineal can spoil scarlet, however.

- If the color is inclined too much towards orange, add 2 teaspoons of alum to the dyebath, reenter the goods, and dye for an additional 30 minutes. Another method is to dissolve 1 to 3 teaspoons of washing soda in 10 gallons of hot water. Work the material in the solution for a few minutes to 1 hour, and then rinse thoroughly.

- If the color is too red, rather than scarlet, add, cautiously, a small amount of good yellow dye to the dyebath, reenter the goods and dye until scarlet.

Cochineal Scarlet Recipe No. 3: Wool

This recipe and Recipe No. 4 are essentially from Hummel (349) and may be taken as representative of the later (1885) premordanting and one-pot methods.

1. Dissolve 1 ounce (8 teaspoons) of cream of tartar in 4 to 6 gallons of distilled, deionized, soft, or rainwater (do *not* use hard water) in a nonreactive vessel. Do *not* use copper or brass unless a piece of metallic tin is added.

2. Dissolve 1/2 ounce (2 1/2 teaspoons) of tin (stannous chloride) in the tartar solution. When all ingredients are dissolved, heat the bath to 120° F, then add the well-scoured, rinsed, damp wool. Work a few minutes and heat the mordant bath up to the simmer. Mordant at this temperature for 60 to 90 minutes.

3. Remove, cool, and rinse or wash thoroughly.

4. Grind 1 1/2 ounces of cochineal insects and simmer them in about 2 quarts of water in a nonreactive vessel for at least 30 minutes. Add the dyeliquor to 4 to 6 gallons of soft water and heat the mixture to 120° F.

5. Add the damp, mordanted wool, heat the vessel to the simmer, and dye for 60 to 75 minutes.

6. Remove, cool, rinse well, and dry.

• If 1/4 ounce of stannous chloride and 1/4 ounce of stannic chloride are used, rather than 1/2 ounce of stannous chloride alone, the color should be slightly yellower and more brilliant.

• If the wool is not scarlet enough, add a small quantity of a yellow dye such as turmeric or black oak bark about two-thirds of the way through the dyeing period.

• Remember to scour wool well in advance. Cochineal will not dye greasy wool.

Cochineal Scarlet Recipe No. 4: Wool

1. Fill a 4- to 6-gallon capacity nonreactive vessel half full of soft water.

2. Dissolve 32 grams (10 teaspoons) of oxalic acid in the water. Then add 1/2 ounce of stannous chloride or 1/4 ounce of stannous chloride and 1/4 of stannic chloride. Stir until dissolved.

3. Grind 1 1/2 ounces of cochineal and add this to the vessel. Stir.

4. Heat to the boil. Boil for 5 to 10 minutes.

5. Remove the heat and fill the vessel the rest of the way up with soft water.

6. Enter the well-scoured, well-rinsed, damp wool.

7. Heat the dyebath to the simmer in about 45 to 60 minutes. Dye at the simmer for 30 minutes longer. Add small amounts of yellow dye if the color is not intense enough or scarlet enough.

8. Remove, cool, wash well with neutral detergent, rinse well, and dry.

• As with all single-bath methods, the dyebath is not exhausted; some dye is lost by combining with the mordant only. This appears as a yellowish flocculent precipitate. This unexhausted bath may be used for additional lots of material with addition of perhaps 2/3 of the original ingredients (the shade may not be identical).

• Hummel's original recipes call for 1 ounce of tin salts. I consider this entirely too much, as it makes the wool harsh. Therefore, I have cut the quantity to 1/2 ounce.

• Less cochineal (1 ounce) may suffice with the premordanting method or if black oak bark liquor is added.

• Potassium oxalate may be used instead of oxalic acid, and instead of stannous chloride and oxalic acid, you may use 23 grams of cream of tartar, 1/4 ounce of stannous chloride, and 1/4 ounce of stannic chloride.

• The single-bath method usually produces the better scarlet.

- Better results may be obtained by using "tin spirits" (see Appendix B) at the rate of 2.5 ounces instead of stannous or stannic chlorides. The reason, according to Hummel, is free hydrochloric acid in the "spirits" mordant. Presence of free acid apparently permits better penetration of thick or hard-spun material before any decomposition of the mordant takes place. Thus, the material will be dyed all the way through, rather than just on the surface. Free hydrochloric acid also neutralizes hard (calcareous) or alkaline water and retains the colored lake in solution longer (this refers to the one-bath method).

- Hummel highly recommends a mixture of stannic and stannous chlorides when using the salts because the stannic helps prevent production of dark or black "tin spots" which sometimes occur in scarlet-dyed piece goods. The spots are, apparently, anhydrous stannic oxide, produced from the hydrate stannous oxide during boiling. Use of oxalic acid also helps to prevent black spots. Use of stannic chloride alone does not produce a brilliant scarlet.

- Cochineal scarlet has good lightfast properties, but the color may be dulled and rendered more bluish by weakly alkaline soaps. Should this occur, a vinegar or weak acetic acid rinse often will restore the bright tone.

Cochineal Scarlet Recipe No. 5: Wool

This recipe is from Napier (398).

1. To 4 to 6 gallons of soft water in a nonreactive vessel add 1.2 ounces of ground cochineal.

2. Add 1 gram of oxalic acid, 9 grams of cream of tartar, 2 1/2 ounces of tin (scarlet) spirits, or 3/8 ounce of stannous chloride and the dyeliquor from 3 to 4 grams of black oak bark.

3. Boil the material for 5 to 10 minutes.

4. Add the well-scoured, rinsed, damp wool, and dye at the simmer for 90 minutes.

5. Remove, cool, wash well, rinse, and dry.

Cochineal Scarlet Recipe No. 6: Wool

This recipe is Poerner's Process (with emendations) as described in *A Practical Treatise* (378–89).

1. Prepare a bath of 4 to 6 gallons of soft water in a nonreactive vessel.

2. Add 14 drachms (7/8 ounce, 25 grams, 8 teaspoons) of cream of tartar. Stir until dissolved.

3. Add 14 drachms (7/8 ounce) of solution of tin (Mordant B, below). Heat to the boil and boil a few minutes.

4. Add the well-scoured, rinsed, and wetted-out wool. Mordant at the simmer for 2 hours.

5. Remove from the mordant bath, let drain, and cool.

6. Prepare the dyebath by dissolving 2 drachms (3.5 grams, 1 teaspoon) cream of tartar in 4 to 6 gallons of soft water in a nonreactive vessel. When dissolved, add 2 ounces of sea salt; (this apparently assists penetration of the dye into the cloth. Regular salt [sodium chloride] may be substituted). Heat to the simmer and add 1 ounce of ground cochineal; stir well and let boil for a few minutes.

7. Let the temperature of the bath drop to the simmer and add slowly and carefully 1 ounce of the solution of tin (Mordant B). Stir for a minute or two.

8. Add the wet, mordanted wool. Keep the temperature of the bath at the simmer. Stir the bath during the dyeing.

9. Remove the wool when the beautiful scarlet has developed.

10. Drain, wash well, rinse, and dry.

• Poerner's and Hellot's scarlet spirits are virtually the same, as follows: Pour, carefully, 8 ounces of concentrated nitric acid into 8 ounces of water (outside, or in a chemical hood). Dissolve 3/4 ounce of sal ammoniac (ammonium chloride) in the diluted acid. Slowly and carefully add 1 ounce of powdered or feathered tin. Add a small bit of tin, then add no more until the first portion is dissolved. Allow the mordant to age for two days before using. Store in a glass stoppered bottle in the dark.

• Mordant "B" has been difficult for me to produce. Powdered tin reacts very slowly in half-concentrated nitric acid. Twelve ounces of concentrated nitric acid slowly poured into 4 ounces of water produces a solution that will react (dissolve) tin.

Cochineal Scarlet Recipe No. 7: Wool

This recipe is from *Dick's Encyclopedia* and is of unknown origin.

1. Scour well, rinse, and leave wool wet.

2. Prepare a dyebath of 4 to 6 gallons of soft water in a nonreactive, or very clean copper vessel.

3. Add .8 ounce (23 grams) of cream of tartar (1/2 ounce tartaric acid may be substituted). Stir until dissolved.

4. Add 1/3 ounce (9 grams) of oxalic acid. Stir until dissolved.

5. Add the dyeliquor from a small amount of good yellow dye. Use a very small amount; more may be added later, if necessary.

6. Add the dyeliquor (or ground insects) from 1 3/4 ounces of cochineal.

7. Heat the dyebath up to 180° F and add 4.5 grams of stannous chloride (previously dissolved in soft water) and 1.5 ounces of scarlet spirits (Appendix B). If scarlet spirits are not available add 11 grams of stannous chloride.

8. Boil the mixture for about 15 minutes, then cool the bath to about 180° F, add the wool, and work well. Heat the bath back up to the simmer or boil. Dye at this temperature for 1 hour. Add additional yellow dye only if the product is not scarlet enough.

9. Remove, cool, and rinse in cold water. If the wool shows any white hair, add, carefully, 1/2 to 3/4 ounce of concentrated hydrochloric acid to the cooling dyebath, reenter the goods for a few minutes, remove, cool, and rinse well again before drying.

Mock Venetian Scarlet: Wool (Liles' Method)

Venetian scarlet preceded gobelin scarlet by centuries and, while less bright and flashy, was considered by many to possess more subtle beauty, and it was definitely more permanent. Venetian scarlet was produced by dyeing wool, premordanted with alum-tartar, with kermes, which is no longer available. Lac was used as well, primarily in India and Persia. Lac is available, at least at times. This recipe uses the more readily available cochineal.

1. Scour well, rinse, and leave wool wet.

2. Premordant for 2 hours at the simmer with alum-tartar, 3 1/2 ounces of alum, 1 ounce of tartar. Rinse well before dyeing.

3. Grind 1 to 1 1/2 ounces of cochineal very well. Simmer the ground bugs for 30 to 45 minutes in 2 to 3 quarts of soft or deionized water in a nonreactive or stainless vessel.

4. Add the dyeliquor to 4 to 6 gallons of deionized or soft water, again in a nonreactive or stainless vessel.

5. Add the wool and heat up to the simmer. Dye at this temperature for 1 1/2 hours. Move the material often and keep a close eye on the color. If the color becomes too blue-red, lift the wool out, and cautiously add small amounts of acid to the dyebath. This should render the dyebath more red or orange-red, and the wool should take on a redder shade (becoming less blue-red). Any acid will work, even

white vinegar. Acetic acid (40%) or half-strength hydrochloric acid works well, also. Add a little acid, stir, and reenter the goods. This procedure is best done during the last 15 to 30 minutes of the dyeing process.

6. Remove material, cool, air, and rinse thoroughly before drying.

Lac Scarlet or Red: Wool

The home of the lac (shellac) insect is India, where the dye was used for centuries. It made its way across overland trade routes and was used in Persia and, to a limited degree, in Europe, particularly prior to the introduction of New World cochineal. Most lac-scarlet recipes from the old literature are similar. The following is from Napier (398).

1. Scour material well, rinse, and leave damp.

2. Prepare a bath of 4 to 6 gallons soft water in a nonreactive vessel.

3. Add 1 ounce of well-ground lac. Stir well. If using lac extract, add a small amount and additional later in the dyeing process, if needed.

4. Add 3 liquid ounces (90 ml) of red (scarlet) spirits (see Appendix B), and 1 teaspoon (5 1/2 grams) of stannous chloride dissolved in 25 ml hydrochloric acid. If no red spirits are available, use 11 grams stannous chloride dissolved in 50 ml hydrochloric acid. Be careful with the acid.

5. Finally, add 8 teaspoons (24 grams) of cream of tartar.

6. Stir the materials, heat to the simmer, and simmer for 15 minutes (190° to 200° F).

7. Add the goods, and dye at the simmer for 1 hour. If the color is not scarlet enough, add a small amount of yellow dye.

8. Remove, cool, rinse, and dry.

Redwood Dyes

The redwoods saw rather extensive use for dyeing from early times, and even occasionally until the end of World War 1 (because of the unavailability of German synthetics). If used alone or as the major dye component, they were usually referred to as "common" or "fancy" reds. When newly made, redwood dyes are quite beautiful, but if exposed to much strong light they fade rather quickly to a reddish brown or near off-white. Frequently, the redwoods were combined with madder or cochineal, to improve brightness, alter the shade, or reduce the required amount of the more expensive fast dye.

The redwoods include several species of trees and shrubs and fall into two basic groups; those containing brasilin (brezilin) as the principal dyestuff, and those containing santalin (santaline). The santalin dyes are the slightly faster of the two, but are more expensive, more difficult to use, and harder to obtain.

The brasilin-producing redwoods include brazilwoods (*pernambuco, bois de Bresil, fernambouc,* queen's wood, redwood, *das Rotholz*). These are species of *Caesalpinia.* Another group of brasilin-producing species, *Haematoxylin brasiletto* et al., came from Nicaragua, Columbia, and Venezuela (Santa Martha wood, peach wood, Nicaragua, hypernick). The brazilwood of the European Middle Ages, *Caesalpinia sappan* (sappanwood), came from India, Malaya, and Ceylon (Adrosko, 25). Thus, the name "brazilwood" was in use in Europe when the early sixteenth-century Spanish explorers discovered similar redwood dye-yielding trees in a South American country, later to be named Brazil.

The second group of redwoods include barwood or camwood, *Baphia nitida,* and sanderswood (saunders, red sanders, santal, sandalwood). Sanderswoods include species of *Pterocarpus,* such as *P. santalinus.* Camwood was imported from the West Coast of Africa, barwood from Sierra Leone, and sanders from India, Ceylon, and the East Indies.

The redwoods were used to dye cotton, linen, wool, and silk; and eighteenth- and early nineteenth-century dyebooks contain many recipes for their use, as does the sixteenth-century *Plichto* by Rosetti. I have tried them, for the most part, only with the cellulosic fibers. Many old and recent dyebooks include recipes for wool. Some of Molony's (1834) recipes are included in Adrosko's book. Redwoods were little used for silk after the introduction of synthetic dyes, but three recipes are listed in Napier's 1875 edition (385).

All of the products of the cotton recipes listed have been subjected to both mild and severe fading tests. The severe test involved subjecting the dyed items to three weeks of full, outside summer sun, from early morning through late afternoon. None survived this test except where the redwood was combined with madder. In this case, it appeared that only the madder remained. This is a very severe test. Any dye, ancient or modern, surviving without noticeable change would be classified as "very fast." The milder test involved placing the dyed items under the banks of fluorescent lights in my laboratory for four weeks. Here, all straight redwoods were altered somewhat, but not too badly. Fluorescent light is far more destructive to dyes than is incandescent, since it contains more ultraviolet. One week under the fluorescent lights proved to be about as destructive as one day of summer sun. Therefore, if the item dyed with the redwoods is to be used, say, in a sweater for late afternoon or evening wear, and stored in the dark, it may be all right. The redwoods would not be suitable for a quilt, coverlet, or afghan, and their best use is in the shading of more permanent reds.

The following recipes are from *Dick's Encyclopedia*, Hummel, *A Practical Treatise,* and Napier. Also, see Madder Red Recipe No. 3 (Common Red) for the madder-brazilwood combination. (All are for 1 pound of material.) The first three recipes should work satisfactorily on silk.

Brazilwood Red Crimson: Cotton or Linen

1. Scour material well, rinse, and leave wet.

2. Mordant with 1 ounce tannin or equivalent dissolved in 4 to 6 gallons of hot water. Work a few minutes, sink the material, and allow to remain for about 12 hours.

3. Remove from the now room-temperature tannin, squeeze, and mordant with basic alum or tin (stannic chloride, sp. gr. 1.01), or tin followed by alum (see tin mordant, or "tin or red spirits," general cotton mordanting instructions, Chapter 3). I suggest mixing 3 ounces of stannic chloride concentrate with 1 gallon of water. Work and soak the material in this for 1 hour, at which time the material should be light yellow. Rinse well.

4. Prepare the dyebath by soaking 4 to 6 ounces finely cut up brazilwood or 3 ounces of brazilwood sawdust in about 2 quarts of hot water. Leave overnight, then simmer or boil for about 1 hour. The overnight soaking is not essential if sawdust is used.

5. Add the brazilwood liquor to 4 to 6 gallons of 120° F water in a copper, brass, or nonreactive or plastic vessel.

6. Add the rinsed, mordanted material to the dyebath, and work thoroughly for about 30 minutes. Dyebath temperature should be between 110° to 130° F.

7. Lift and add 1/2 ounce of concentrated tin solution or red spirits or alum mordant, place the goods back in the bath, and work for another 15 minutes.

8. Remove, squeeze, rinse, and dry.

Brazilwood Common Red: Cotton or Linen

This recipe is exactly the same as Brazilwood Red Crimson (above), except that the dyebath is made up of 3 to 4 ounces of brazilwood chips and 1 ounce of old fustic or 2 ounces of brazilwood sawdust and 1 ounce of old fustic.

Brazilwood Common Crimson: Cotton or Linen

This recipe is also exactly the same as the first, Red Crimson, recipe except that the dyebath is made up of 3 to 4 ounces of brazilwood chips or 2 ounces of brazilwood sawdust and 1 ounce of logwood.

This recipe produces a bluer crimson than the first recipe, which is a redder crimson. In-between shades may be obtained by decreasing the amount of logwood. Prettier colors are obtained using tin rather than alum mordant.

Barwood or Sandalwood (Mock Turkey Red): Cotton or Linen

The barwood, camwood, or sandalwoods are so hard, and the dyestuff so little soluble, that they must be used in sawdust form. The sawdust is added directly to the dyebath. This works satisfactorily with cotton yarn at least, but the old literature indicates problems with silk. Because these dyewoods, particularly barwood, are sometimes unavailable, I have had the opportunity to use them only three times.

1. Scour the cotton well, rinse, and leave wet.
2. Mordant with 2/3 ounce of tannin in 4 to 6 gallons of hot water. Add 10 to 15 ml (1/2 ounce) concentrated sulfuric acid. Mordant for 12 hours.
3. Remove, squeeze, and work in tin spirits or stannic chloride for 30 minutes.
4. Squeeze out and rinse.
5. Put 1 to 2 pounds of barwood or 2 pounds of sandalwood sawdust into a nonreactive vessel with 4 to 6 gallons of water. Heat. When the dyebath temperature reaches about 180° F, enter the cotton and work it among the wood for about 20 to 40 minutes. The dyebath temperature may be permitted to reach the simmer or boil.
6. Remove, squeeze, rinse well, and dry.
• If the dyed material starts to change from red to a reddish-

brown, remove and finish off even if only 20 minutes have elapsed. Prolonged heating dulls this color, and a brighter red is obtained by using more dyestuff and a shorter time in the dyebath.

Safflower

This is an ancient pink and rose dye of India, the far East, and Egypt, where it grew indigenously. Botanically, it is *Carthamus tinctorius,* an annual thistle. In the Middle Ages and later it was cultivated in Europe, and it can be grown in North America. As Adrosko points out (30), it also went under the names of "carthamus" and "bastard saffron." Safflower (the flower heads only) contains three dyes—two rather worthless yellows and one good, though fairly fugitive, clear pink or red (Napier, 290). The red is suitable only for cotton and silk, which it dyes substantively. The colors fairly smack of our notion of ancient Indian costume pinks and roses, and in shade they are unique and different from other traditional pinks and reds. According to Adrosko the most significant use of this dye by Westerners was in dyeing cotton tapes for legal documents—the original "red tape."

Though they must be kept from strong sunlight, safflower pinks and roses are most beautiful on cotton, and it is possible to achieve stunning lilacs and lavenders by carefully overdyeing Prussian blue with safflower. Though safflower may be grown, I sometimes obtain the dried flowers from health food stores.

Safflower Pinks, Rose, and Crimson: Cotton and Linen

The shades listed above are dyed identically, the difference depending upon the amount of safflower used. Napier suggests approximately 3.5 ounces of flowers for pink, 7 ounces for rose, and 11 ounces for crimson. *A Practical Treatise* suggests 16 ounces of flowers for deep red. I have obtained excellent results with these quantities.

1. First the worthless yellow dye must be washed away. Place 1 pound or less of dried flowers in a 2-gallon (or greater), wide-mouth glass jug. Fill the jug with cold water and stir well. Stretch and tie a fine mesh nylon gauze (such as from old hose) over the mouth to serve as a fine sieve. Whenever convenient, pour away the water and refill with cold water. Continue doing this until virtually all of the yellow dye is washed away. Safflower contains a tremendous amount of yellow dye, so don't give up too soon.

2. When the flowers are free of yellow, dissolve 1 ounce of pearlash or washing soda in 2 gallons of cold water and add this to the jug with 1 pound of damp flowers. Reduce proportionally for other quantities. Too much alkali will ruin the dye. Stir and allow to stand for 5 to 6 hours, then pour off the extracted dye. The alkali extracts the red dye as well as the second yellow dye. Add a little water to the flowers, stir slightly, and add this to the other dyeliquor. Discard the spent flowers. The dyeliquor spoils fairly rapidly. Store it in the cold for a day or two if necessary; otherwise use it immediately.

3. Add enough additional cold water to the dyeliquor in a nonreactive or plastic vessel to give a 4 to 6 gallon total volume.

4. Work the well-scoured, damp cotton in the cold dyebath for about 5 minutes, then lift it out and add, slowly and carefully, about 3 ounces concentrated sulfuric acid or equivalent of weaker acid. Stir, add the goods again, work a while, sink the material, and allow to remain for about 1 hour. Use 4 ounces of acid for the crimson.

5. Remove from the dyebath, squeeze, and rinse in 3 separate waters. Dissolve 1 to 2 level teaspoons of cream of tartar in the last water (dissolve in hot water and add to the cold rinse water).

6. Remove, squeeze, and dry in the shade.

• The sulfuric acid may be added to the dyebath before introduction of the cotton. Indeed, this is the way it *must* be done when over-dyeing Prussian blue with safflower for lilac because alkaline solutions decompose Prussian blue.

Safflower Red: Silk

Safflower dyes silk as well as cotton, but the second worthless yellow must be removed first. This yellow is not taken up by cotton, but it is by silk. To circumvent this problem, proceed as follows: prepare the dyebath for cotton (acid added) and add a quantity of scoured, bulky, open-weave cotton. The cotton will take up the red, but not the yellow. Now wash the cotton well in cold water to wash free all of the yellow. Next place the material in 4 to 6 gallons of water in which has been dissolved 1 ounce of washing soda or pearlash. Work the cotton for 15 to 30 minutes. The red dye will be discharged back into the dyebath. Remove the cotton, add acid, acidifying the bath, and add the scoured wet silk. Dye the same as for cotton.

Safflower red and scarlet may be obtained by dyeing first with annotto (a deeper orange for scarlet), followed by 9 to 11 ounces of safflower. (See Cochineal Scarlet Recipe No. 2: Silk for preparation of annotto).

American Hopi Indian Red Dyes

Unfortunately, I have not had the opportunity to experiment with any of the Hopi dyes. However, the interested reader is referred to *Hopi Dyes*, by Mary-Russell Ferrell Colton. Included are reds for wool, cotton, and basketry material. These are the old, traditional recipes.

6 Green Dyes

Prior to the advent of synthetic dyes, the majority of good clear greens, oranges, and purples were produced by overdyeing one primary color with another. Thus, the very early greens resulted from overdyeing indigo with a clear yellow dye or vice versa. Some of these greens were satisfactory for long exposure to light, and others were not. Weld, old fustic, and black oak bark yellows were among the best, particularly if small amounts of copper were used in addition to alum in the mordanting procedure. In India, turmeric was often used for yellow early on, as was young fustic (Venetian sumac) and Persian berries in Europe. Because these yellows are fugitive, many old fabrics show far more blue and less green than was originally present. The yellows have mostly faded away, leaving the indigo blue. On the other hand, it is most gratifying to see greens remaining in medieval tapestries and similar old fabrics where the yellow dye was weld.

Off-greens and olives were obtained by premordanting with alum, dyeing a yellow, and aftermordanting with iron near the end of the dyeing procedure, or dyeing a yellow with iron mordant as a one-pot process.

Reasonably fast natural yellow-logwood greens were in use, incorporating copper mordants, starting probably in the eighteenth century. Prussian blue was produced in France in 1749, and it produced lightfast greens, particularly with weld, black oak bark, and old fustic. Scheele's green (blue-stone sage, green arsenic sage, or arsenic sage) appeared about 1770, and while it was lightfast on cotton, linen, and paper, it was extremely poisonous. Napier complained bitterly in 1875 that the dye was still being used and that it poisoned the maker, the winders of yarn dyed with it, and the person using the dyed article. It was particularly bad if used on wallpaper in bedrooms since the occupant usually spent 7 to 9 hours per day in such a room. Very good evidence indicates that Napoleon died of arsenic poisoning while in exile on the island of St. Helena and that the poisoning may have been accidental. Recent evidence indicates that servants in the house may also have suffered arsenic poisoning, and that the rooms were probably covered with blue-stone sage wallpaper. Because of the dye's extremely poisonous nature I have decided not to list a recipe for its production. Should a tiny amount need to be made for historical purposes, a recipe exists in *A Practical Treatise,* but the material should be produced only by a competent chemist.

Acid extract of indigo (indigo sulfonate) appeared around 1740–50 and was much used in conjunction with natural yellows for "Saxon" greens, particularly on wool and silk. Around 1840 the very clear, lightfast and washfast (and poisonous) lead chromate yellow appeared. This dye was used as an overdye with Prussian blue or indigo to produce fast greens on cotton, linen, and paper. It was probably the most used green on cellulosic materials until about 1910. Since it withstands exposure so well, it was much used for outdoor green awning material and, according to Pellew (1912), to color our "greenback dollars" from about 1850–1900.

All recipes are for 1 pound of material.

Indigo and Yellow Dye Combination Green: Cotton and Linen

For best results, the indigo should be dyed first. In most cases, the article will be more evenly dyed with the indigo prior to mordanting, and level dyeing is essential for good greens. If too dark a yellow overdye results, producing an unwanted yellow-green, the material can then go back into the indigo. In order to avoid too dark a yellow, use a relatively weak yellow dyebath, adding more until the right shade of green is obtained, or remove the material from the yellow bath when the shade is right.

1. Scour material well, rinse, and dye a medium or medium-dark shade of indigo (see Blue chapter). Build the shade up to the desired level by redipping at least 3 times. Make certain that the indigo is well penetrated into the yarn or piece goods. Allow the material to air overnight or longer before rinsing out. A slightly darker shade of blue than considered proper is desirable because a little will be removed in the subsequent mordanting and overdyeing, and the surface indigo will light-fade slightly.

2. Mordant once or twice with aluminum acetate or twice with Basic Alum Mordant No. 2. Permit adequate ageing time following mordanting.

3. Treat well with fixing solution or soak and rinse several times.

4. Prepare a good yellow dye concentrate (see Yellow chapter). Add 1/3 to 1/2 of the yellow concentrate to 4 to 6 gallons of soft water, unless you are using black oak bark or weld. Use a nonreactive, copper, or brass vessel. Suggested yellow dyes include old fustic, Osage orange, goldenrod, arsemart, peach leaves, or Queen Anne's lace. Black oak bark does not seem to produce the prettiest greens on cotton.

5. Heat the dyebath to 120° to 160° F, then add the wetted-out material. Keep working the material. It should take on a nice green shade in 10 to 20 minutes. Lift and add additional yellow dye concentrate if the shade is not dark enough in 30 minutes.

6. If the yellow dye is not taken up in 30 minutes or is taken up unevenly, lift and add about 2 teaspoons of alum, previously dissolved in 1 quart hot water. Reenter the goods and work well. Additional yellow dye may be needed following this procedure.

7. When the shade starts looking correct, lift and add about 1 1/2 teaspoons of copper sulfate, previously dissolved in 1 quart of hot water. Stir, reenter the goods, and dye for about 10 additional minutes.

8. Remove, wash well, and dry.

Indigo and Yellow Dye Combination Green: Silk

Indigo-yellow dye silk greens are produced using the same method as for cotton except for the following:

1. Do not permit the silk to dry completely following the indigo treatments. Instead, place the silk items in water 30 or so minutes after the final dip in the indigo vat.

2. Following dyeing with indigo, mordant the silk twice with either aluminum sulfate or aluminum acetate.

3. After the mordanted silk is dry, rinse it well with water.

4. Omit aftermordanting with copper sulfate; instead add 2 to 3 ounces of the alum mordant approximately 10 minutes before removal from the yellow dyebath. Old fustic works well on silk.

Indigo and Yellow Dye Combination Green: Wool

Indigo-yellow dye combinations are done exactly the same as for cotton except for the following:

• Remember that the same strength indigo vat will give darker shades on wool than on cotton or silk. If all three fiber types are to be dyed in the same indigo-hydrosulfite or thiourea dioxide vat, dye the cotton items first, then the silk, and finally the wool.

- Mordant with alum-tartar following the indigo dyeing.
- The yellow dyebath for wool can run about 20° F higher than for cotton or silk.

Olive Green: Cotton or Linen

1. Scour material well, rinse thoroughly, and leave wet.
2. Prepare a dyebath by soaking and boiling 1 ounce fine old fustic sawdust or 2 to 3 ounces of fustic chips.
3. Add the fustic concentrate to 4 to 6 gallons of hot water (130° to 180° F).
4. Add 12 grams (3 1/2 teaspoons) of copperas previously dissolved in 1 quart of hot water.
5. Add the cotton or linen and work well for about 30 minutes. Keep the material moving rather frequently so that the iron mordants evenly.
6. Remove, wash well with detergent, rinse, and dry.

Yellow Dye Off-Green: Wool

1. Dye the wool yellow, using alum-tartar premordanted wool (see Yellow chapter).
2. Near the end of the dyeing procedure, lift wool out and add about 1 teaspoon of copperas previously dissolved in 1 quart of hot water. Stir and reenter the wool. Work well and continue dyeing for about 10 additional minutes.
3. Remove, wash well with detergent, rinse, and dry.

- This procedure will often produce an olive or grey-green, sometimes a tan. Also, greens so produced may fade to an off-brown in time.

Logwood-Fustic Gray-Green: Cotton or Linen

This recipe produces a nice, relatively light grey-green for me. However, it is listed as a brown recipe in Bronson.

1. Scour goods well, rinse, and leave wet.
2. Dye the cotton olive green (Olive Green Cotton Recipe).
3. Prepare a dyebath concentrate from 1 ounce of logwood sawdust or 2 ounces of logwood chips.
4. Add the logwood concentrate to 4 to 6 gallons of hot water (140°

to 160° F). Add 6 grams (1 1/2 teaspoons) of copper sulfate previously dissolved in 1 quart of hot water.

5. Add the olive dyed cotton and dye for about 30 minutes.

6. Remove, wash with detergent, rinse well, and dry.

Logwood-Fustic Dark Amish Green: Cotton or Linen

1. Scour material well, rinse, and leave damp.

2. Mordant twice with Basic Alum Mordant No. 2 or aluminum acetate. Allow adequate ageing time following each mordanting.

3. Treat well with fixing solution or soak and rinse several times.

4. Prepare an old fustic concentrate from 1 ounce of fine fustic sawdust or 2 to 3 ounces of chips. Add this to 4 to 6 gallons of hot water (120° to 160° F) in a nonreactive, copper, or brass vessel.

5. Add the cotton and work well since the dye will be taken up very quickly. The cotton will be a fairly dark golden yellow in about 15 minutes.

6. Lift material and add about 7 grams (1 1/2 teaspoons) of copper sulfate previously dissolved in 1 quart hot water.

7. Reenter the goods and dye for 10 additional minutes.

8. Remove, squeeze, and air.

9. Dissolve 21 grams (5 teaspoons) of copper sulfate in 4 to 6 gallons of hot water (140° to 180° F).

10. Add the olive-yellow dyed cotton and mordant for about 30 minutes. Work the material several times during the mordanting. Remove, squeeze, and air.

11. Prepare a logwood concentrate from 6 ounces of logwood chips or 2 to 3 ounces of sawdust. Boil or simmer the logwood in about 2 gallons of water for 45 to 60 minutes.

12. Add the logwood liquor to enough water to make a dyebath of 4 to 6 gallons. Use a nonreactive, copper, or brass vessel. Keep the dyebath temperature at 150° to 180° F.

13. Add the cotton and dye for about 30 minutes.

14. If yellow-green, lift the material and add 12 grams (2 1/2 teaspoons) of soda ash or pearlash. Stir well. Reenter the goods and dye for 15 to 20 additional minutes. This will intensify the blue, producing a deeper green.

15. Remove, squeeze, wash well with detergent, rinse, and dry.

• Osage orange or other good yellow dyes may be substituted for old fustic.

Prussian Blue–Fustic Green: Cotton or Linen

Nineteenth-century dye manuals contain numerous Prussian blue–old fustic recipes. This recipe and the following one are typical and work quite well.

1. Dye goods a medium shade of Prussian blue (see Blue chapter).
2. Mordant, carefully, in medium-strength aluminum acetate for at least 1 hour. Squeeze out carefully and dry overnight. (Basic Alum Mordant No. 2 may be substituted. If so, mordant twice.)
3. Rinse material carefully in warm water.
4. Work carefully for 30 minutes in a dyebath of fustic at 100° F (make a fustic concentrate from 4 to 5 ounces of fustic chips or 2 ounces of fine fustic sawdust and add to 4 to 6 gallons of water).
5. Lift the goods and add 6 grams (2 teaspoons) of alum previously dissolved in 1 quart of water. Stir and reenter the goods. Dye for an additional 10 minutes.
6. Remove, squeeze, and dry.

• A number of shades may be produced by varying the depth of Prussian blue and the quantity of fustic.

• Black oak bark or Osage orange may be substituted for the fustic.

Olive or Bottle Green: Cotton or Linen

1. Dye goods a medium or sky-blue shade of Prussian blue (see Blue chapter).
2. Mordant for at least 1 hour with medium-strength aluminum acetate or twice with Basic Alum Mordant No. 2, allowing adequate drying time between mordanting sessions.
3. Rinse thoroughly several times.
4. Prepare a 4 to 6 gallon dyebath using 4 1/2 ounces of fustic chips or 2 ounces of fine fustic sawdust and 1 1/2 ounces of sumac leaves or 9 grams (3 teaspoons) of tannic acid. Heat the dyebath to about 120° F.
5. Add the goods and dye for 30 minutes, then lift them out and add 20 to 25 ml of "iron liquor" (Appendix B) or 1 teaspoon of copperas previously dissolved in 1 quart hot water.
6. Reenter the goods and dye for 15 minutes. Work the goods well.
7. Remove, squeeze, and wash in a solution made by dissolving

6 grams (1 1/2 to 2 teaspoons) of alum in 4 to 6 gallons of water at 100° F.

8. Remove, squeeze, and dry.

Prussian Blue–Fustic–Logwood Kelly Green: Silk

The older literature rarely suggests using Prussian blue–based greens on silk. My own experimentation resulted in this recipe, though, and it produces a most striking and beautiful Kelly green.

1. Dye silk a medium Prussian blue (see Blue chapter). I generally use ferric sulfate and yellow prussiate with nitric acid, but copperas or "iron liquor" and sulfuric acid is satisfactory. Do not throw out the acidified prussiate.

2. Mordant carefully with aluminum sulfate or aluminum acetate.

3. Following drying, rinse well in warm water.

4. Prepare a dyebath using the dyeliquor from 4 1/2 ounces of fustic chips or 1 ounce of fustic sawdust and 1 1/2 ounces of logwood chips or 3/4 ounce of logwood sawdust. Fill the dye vessel to 4 to 6 gallons and heat to 120° to 160° F.

5. Add the damp silk and work well for 15 or 20 minutes.

6. Lift and add 1 ounce alum mordant. Reenter the goods and dye for an additional 10 minutes.

7. At this point, if Kelly green is desired, work the material for several minutes in the acidified yellow prussiate. It is at this point that the shade changes to Kelly green.

8. Remove, squeeze, rinse, and dry.

Chrome Green: Cotton and Linen

Chrome greens were much used from about 1840–1910 because of their extreme fastness. They may be produced by overdyeing either indigo or Prussian blue with chrome yellow. The indigo or Prussian blue must be dyed first. Bear in mind that chrome yellow is poisonous. The dyed material should be well washed with neutral detergent and rinsed thoroughly. Make certain that the lead nitrate or acetate and chrome solutions are poured into a hole in the ground away from water sources, etc.

1. Dye goods an indigo or Prussian blue approximately of the depth of green desired (see Blue chapter).

2. Overdye the blue with chrome yellow (see Yellow chapter).

• Make certain to dye the blue dark enough and not to make the yellow too dark. Once the yellow is dyed, the material cannot receive additional blue dye.

Indigo Sulfate–Yellow Dye Green: Silk

Many recipes exist for producing silk greens using indigo extract, and they can be quite beautiful. However, they should not be subjected to large amounts of sunlight or the item may have to be re-dyed. Neither are they as washfast as vat indigo or Prussian blue-greens. The greens produced are of different shades than those produced by vatted indigo or Prussian blue.

1. Scour well or wet out the silk thoroughly.
2. Prepare an alum or aluminum sulfate solution using 8 ounces of alum per gallon of water. Mordant the silk cold for 1 to 2 hours. (Aluminum acetate also works very well.)
3. Remove, squeeze out, and rinse thoroughly in warm water.
4. Prepare a dyeliquor from 12 ounces of fustic or Osage orange wood chips or 4 to 6 ounces of fine sawdust.
5. Add the dyeliquor to 4 to 6 gallons of water at approximately 100° to 120° F.
6. Add the damp silk; work and dye the material for about 30 minutes. If the silk is lightly or unevenly dyed, add 4 ounces of the alum mordant and work for several additional minutes.
7. Lift the silk and add about 1/2 to 1 liquid ounce of indigo extract. Stir, reenter the silk and dye for 30 additional minutes. Heat the dyebath only if the green does not develop. Avoid temperatures higher than 130° F. If the shade is too yellow-green to suit, add additional indigo extract. If it is too blue-green, add more fustic. (Addition of dyeliquor from 1 ounce of logwood sawdust to the fustic will produce a bottle green.)
8. Remove, squeeze, rinse well, and dry.

Bottle Green: Silk

1. Scour or wet out the material well and leave damp.
2. Prepare a mordant by dissolving 8 ounces of alum or aluminum sulfate and 3 1/2 ounces of copperas in 3 gallons of water.
3. Add the silk and mordant for 1 hour. Move or work the material frequently for even mordanting.
4. Remove, squeeze, and rinse in warm water.
5. Prepare a dyeliquor concentrate from 12 ounces of fustic chips

or 6 ounces of fine fustic sawdust. (Osage orange may be substituted for fustic.) Add this to sufficient water to make up a 4 to 6 gallon dyebath.

6. Add the mordanted silk and dye for 30 minutes at approximately 120° F.

7. Lift the material and add 1/2 to 1 ounce of indigo extract. Stir and reenter the goods. Dye for an additional 20 to 30 minutes. Dyebath temperature should be about 120° F. Dyeing alum-mordanted silk at high temperature destroys the lustre of the silk.

8. Remove, squeeze, rinse well, and dry.

Saxon Greens: Wool

Saxon greens on wool are most striking and beautiful, particularly if old fustic or Osage orange is used as the yellow dye. Bear in mind not to subject them to any more sunlight than necessary. Many shades are possible, depending upon the relative amounts of indigo extract and yellow dye used.

1. Scour wool well, rinse, and mordant with alum-tartar for about 1 hour. Remove, squeeze, and air.

2. Prepare a dyebath by adding 1/4 ounce of indigo extract to the mordant bath, stir, and reenter the wool. Work well for about 5 to 10 minutes. Bath temperature should be 170° to 190° F.

3. Lift and add another 1/4 to 3/4 ounce of indigo extract, stir, and reenter the wool. Dye for an additional 10 to 15 minutes.

4. Remove the wool, squeeze, and air again.

5. Remove half of the dyeliquor and discard. Add cold dyeliquor, of the same volume removed, produced by 1 to 2 ounces of fine fustic sawdust or 3 to 4 ounces fustic chips. Temperature of the dyebath should be 130° to 150° F.

6. Reenter the wool and dye for about 30 minutes. Do not permit the dyebath temperature to exceed 170° F.

7. Remove, squeeze, cool, wash with detergent, rinse, and dry.

• If the shade does not suit, additional indigo extract and/or fustic may be added. Better results are obtained by dyeing the blue at higher temperature than the yellow. Osage orange may be substituted for the fustic.

"Bancroft's Mordant" Green: Wool

This recipe approximates Bancroft's earlier method, with stannous chloride being substituted for "tin spirits."

1. Shred 1 1/2 ounces of black oak bark in 2 to 3 gallons of water in a nonreactive, copper, or brass vessel.

2. Heat to the simmer for about 1 hour.

3. Add 18 grams (3 1/2 teaspoons) of alum. Stir until dissolved.

4. Add 9 grams (1 1/2 teaspoons) of stannous chloride. Stir until dissolved.

5. Boil the mixture for about 5 minutes, then add 2 to 3 gallons of cold water.

6. Add 1 to 3 ounces of indigo extract and stir.

7. Add the well-scoured damp wool and heat the dyebath to 180° to 190° F. Work well after first introducing the wool. Dye at the simmer for about 30 minutes. Additional bark liquor or extract may be added if the color does not suit.

8. Remove, squeeze, cool, wash well, and dry.

Olive Colors: Wool

Many yellow dyes yield olive colors if premordanted with copper-tartar. If the olive is not dark enough, the material may be after-mordanted with copperas. Old fustic and Osage orange are suggested yellow dyes.

Copper Green: Wool

Premordanting with copper-tartar gives a nice sea green with most wools if the copper sulfate is of high quality.

7 Purple Dyes

The most famous purple from about 1500 B.c. to the Middle Ages was Tyrian or "Royal" purple. It is assumed that this dye was discovered and perfected by the Phoenicians. According to Brunello (57), the name "Canaan" (present-day Palestine and Syria) was derived from the Hebrew word for "purple." By Roman times, Carthage, Ventura, and Cadiz, as well as Tyre and Sidon became centers of the Tyrian purple dyeing industry. According to Pellew (1928, p. 27), the cost of dyed material in the early Roman era was around the equivalent of $350 or more per pound in gold. Thus the term "born to the purple": only the most wealthy could afford it or, in some cases, were allowed to wear it.

Snails of the genera *Murex* and *Purpura* (Mediterranean Sea) and *Purpura* (Atlantic and Pacific oceans off the shores of Nicaragua and Mexico) produce a secretion that contains the dye. The *Murex* snails are carnivorous and were dredged up and killed in order to extract the dye. By the Middle Ages they were so depleted that the industry died out. The *Purpura* snails of Mexico and Nicaragua, on the other hand, are sessile on rocks and may be picked up, milked of secretion, and placed back in the water. This was, and is, the method used by the Nicaraguans and Mexicans (this method of dye extraction is still practiced by a few Mexican natives in Oaxaca state, Mexico [Wipplinger, 1985].

Again, according to Pellew, in about 1908 a German dyechemist, Dr. Friedlander, spent the summer in Naples and collected approximately 12,000 *Murex* snails for dye extraction. From this quantity of snails he was able to extract about three-fourths of a gram of pure dye. Upon analysis, the dye proved to be 6, 6' dibromoindigo. Part of Friedlander's interest was in determining whether Tyrian purple was identical to thio-indigo red B, a synthetic indigo derivative that he had recently produced. The dye was not the same, and Tyrian purple was synthesized and used only for a short while. Synthetic violet vat dyes, equal in fastness, and superior in brilliance, were already available. Pellew (1928, p. 30), mentions that some excellent examples of cloth dyed with Tyrian purple from the Egyptian tombs reside in the Metropolitan Museum of Art. In 1971, another German chemist, Helmut Schweppe, worked with both synthetic and natural Tyrian purple. Dyed yarn specimens shown in his article (*BASF Review*, 1976), in color, indicate that several shades of Tyrian purple were produced. In actuality, all of the species of snails produce a mixture of both dibromoindigo and normal indigo (indigotin). The greater the percentage of the brominated derivative, the greater the tendency towards a reddish purple or what was termed scarlet. Pliny the Elder apparently described Tyrian purple as having the color of freshly coagulated blood, which is a reddish purple.

In the Middle Ages, archil lichen dyes were used for purple, though they were rather fugitive compared to Tyrian purple. Even in very early times, the expensive Tyrian purple was "stretched" by addition of the lichen dyes. Also, kermes or lac overdyed with woad or indigo produced good, permanent purples on wool and silk. A most pleasing reddish brown-purple was developed by the East Indians for cotton by using madder red overdyed with indigo. This was eventually called "Egyptian purple." I can also produce this sort of color by dyeing cotton with madder or alizarin after premordanting with iron-contaminated alum. The early Egyptians probably did the same. Early, good Indian purples were also produced on cotton with tannin, iron, and madder. Following introduction of cochineal and logwood into Europe, a much greater array of purples was possible, and by about 1760 safflower lilac (Prussian blue overdyed with safflower on cotton) appeared.

All recipes are for 1 pound of material.

Amish Madder Purple: Cotton or Linen

The best recipe ideas I have discovered for madder or alizarin purple are from Hummel (452–53). These purples and clarets result from premordanting with oil or tannin or both, followed by an iron or iron and alum mordant. The medium-to-dark shades are the prettiest: a nice looking old-style Amish purple. The light shades may tend towards a brownish purple. The following procedure is fairly simple and works well.

1. Scour well, squeeze, and leave the material damp.
2. Mordant at the rate of 1 to 1 1/2 ounces of tannin or equivalent. Use full 1 1/2 ounces if a darker purple is desired. Remove from the tannin after 10 to 12 hours.

3. Squeeze out well and mordant with ferrous sulfate, ferric sulfate, acetate of iron, or nitrate of iron. Depth of shade will depend, in part, upon amount of iron used. A suggested quantity is 2 to 4 teaspoons of copperas or ferric sulfate dissolved in 4 to 6 gallons of warm water. Any kind of vessel, including plastic, may be used. Work the material for about 30 minutes in the iron, squeeze out well, and air. Check for level mordanting. If unlevel or light, work the material for another 15 minutes, squeeze, and air again. Repeat again if necessary, squeeze out well, and rinse thoroughly. (Iron mordanting can be tricky. It is best to keep turning the material to avoid spotting or streaking.)

4. Prepare a dyebath using an excess of madder or alizarin. This is necessary because a dull color will result unless all of the mordant is combined with dye. Make up the dyebath on the basis of 8 ounces of madder or 5 grams of alizarin (for production of the dyebath, see madder red recipes). Make certain that a teaspoonful of calcium acetate or chalk is added unless the water contains quite a bit of calcium salts.

5. Add the well-rinsed material and work for a few minutes at room temperature. Then heat the dyebath slowly up to about 160° to 180° F and keep it at that temperature for about 1 hour.

6. Remove, squeeze, rinse or wash thoroughly, and dry.

Pennsylvania German Madder or Alizarin Violet: Cotton or Linen

1. Scour goods well and mordant with 1 ounce of tannin dissolved in 4 to 6 gallons of hot water. Let the material remain in the tannin a minimum of 6 hours.

2. Remove and squeeze out well.

3. Mix 8 ounces of black mordant and 1 ounce of aluminum acetate with enough water to cover the goods. Mordant with this at room temperature for 1 to 2 hours. (See Appendix B for preparation of black mordant.)

4. Prepare a dyebath using 1 pound of madder or 10 grams of alizarin in at least 5 to 6 gallons of water. Use any kind of vessel capable of being heated.

5. Wash the mordanted goods thoroughly, or better yet, treat with fixing solution.

6. Enter the wet goods and work for several minutes at room temperature. Then heat the dyebath slowly up to the simmer. Dye at the simmer for about 1 hour.

7. Remove, squeeze, wash well, and dry.

• This dye may be produced by premordanting first with tannin, then with copperas, and then with weak basic alum. However, the item must receive far more iron than alum; otherwise a brown will result.

• The original recipe (Kuder) does not call for use of tannin. If omitted, aluminum acetate should be used.

Amish Logwood Purple: Cotton or Linen

This recipe is a modification from Bronson (138–39). I had thought it might fade rapidly, but it has held up much better than expected, and the fading is true; that is, it just becomes lighter with time. The shades produced are similar to those achieved with madder or alizarin, tannin and iron (Amish Madder Purple).

1. Scour well, rinse, squeeze, and leave material damp.
2. Prepare a tannin mordant bath by dissolving 2 ounces of tannic acid or equivalent in 4 to 6 gallons of hot (120° to 140° F) water.
3. Work the material in the tannin for a few minutes, then sink and leave overnight.
4. Remove, squeeze well, and leave damp or dry.
5. Prepare a logwood dyebath by boiling 10 ounces of logwood chips or 5 ounces of logwood sawdust in 1 to 2 gallons of water for about 1 hour. When cool enough to handle, pour the dyeliquor away from the sawdust or chips.
6. Place half of the logwood liquor in a vessel capable of being heated and add enough hot water to make a total volume of 4 to 6 gallons. The temperature of this dyebath should be 120° to 160° F.
7. Add the cotton, work and dye for 20 minutes.
8. Remove, squeeze, and air.
9. Dissolve 10 grams of alum (2 teaspoons) in a pint of hot water, then add 3 to 4 grams (1 teaspoon) of verdigris or copper sulfate. Stir until dissolved. Add these materials to the dyebath.
10. Reenter the aired cotton; work and dye for another 20 minutes.
11. Remove, squeeze, and air as before.
12. Add half of the remainder of the hot logwood liquor concentrate to the dyebath, reenter the aired cotton, and dye for another 20 minutes.
13. Remove, squeeze out, and air the material.
14. Pour away the dyebath and put the remainder of the logwood concentrate in the vessel; add enough hot water to make 4 to 6 gal-

lons. Dissolve 5 grams (1 teaspoon) of alum in the dyebath and, when it has dissolved, reenter the goods for another 20 minutes of dyeing.

15. Remove, squeeze, air, rinse, and dry.

• This recipe exhibits the very important trick of airing (oxidation), much used by the early dyers. The very small amount of copper is undoubtedly of extreme importance, not only for final color, but also for colorfastness.

Logwood Purple: Cotton or Linen

The old dye manuals abound with recipes comparable to this one. However, I can recommend it only for historical purposes because it is not lightfast and it is quite expensive to produce. The dye fades from a very nice reddish purple to a rather ugly off-brown.

1. Scour material well, squeeze out, and leave damp.

2. Mordant with 23 grams (3/4 ounce) of tannic acid or equivalent. Work the material in the hot tannin, then sink the material and allow it to remain overnight.

3. Remove, squeeze well, and work in "red spirits" or a stannic chloride solution for 1 hour. (To produce this mordant from stannic chloride, dissolve 3 ounces $SnCl_4$ in 1 gallon of water.)

4. Remove from the tin solution and wash well in cold water.

5. Prepare a logwood liquor concentrate by boiling 5 ounces of chips or 2 to 3 ounces of sawdust in 1 to 2 gallons of water for 1 hour. Decant the liquor into a nonreactive, brass, or copper vessel, and bring the volume up to 4 to 6 gallons with hot water.

6. Add the cotton and dye the material for 30 minutes at 130° to 150° F.

7. At the end of 30 minutes, lift the material and add 4 ounces of the tin solution or red spirits, or 1 1/2 teaspoons (6 grams) of potassium alum. Stir well, and reenter the goods for another 10 minutes.

8. Remove, squeeze, rinse well with cold water, and dry.

• For a bluer purple, use more logwood and less tannin, and, after lifting, add alum instead of the red spirits.

Logwood–Prussian Blue Lilac (Puce): Cotton or Linen

1. Dye goods a medium shade of Prussian blue (see Blue chapter).

2. Prepare a logwood dyebath by boiling 1 ounce of logwood sawdust or 2 ounces of logwood chips in 4 to 6 gallons of water. Remove the chips or pour away from the sawdust, and when the temperature

has dropped to 120° to 140° F enter the wetted-out goods and work for about 15 minutes.

3. Lift the material and add 12 grams (2 1/2 teaspoons) of alum. When the alum is dissolved, reenter the goods and dye for another 10 to 15 minutes.

4. Remove, squeeze, rinse well in cold water, and dry.

• I have not tested this recipe for long-term lightfastness. However, the Prussian blue is quite lightfast.

Safflower Lilac: Cotton or Linen

Safflower lilac, newly made, is extremely beautiful. However, do not expose the material to strong light or sunlight for great lengths of time. Again, the safflower appears to hold up longer in conjunction with the Prussian blue than alone. Recipes such as the following remained in dye manuals as late as 1885.

1. Produce a medium shade of Prussian blue (see Blue chapter) on cotton or linen. (If piece goods, make certain that the Prussian blue is dyed quite level.)

2. Overdye the wet Prussian blue–dyed material with a safflower pink using about 5 ounces of safflowers (see Red chapter for safflower recipes). In this case, the acid must be added to the safflower dye before immersing the Prussian blue–dyed material because alkaline solutions discharge Prussian blue back to iron buff.

3. Remove, squeeze, rinse in cold water, and dry the material in the shade.

• A number of different hues are possible by varying the shade of Prussian blue, safflower, or both.

• Another way of producing this dye is by overdyeing safflower with Prussian blue. This method has not worked as well for me, however, and it requires more safflower.

Egyptian Purple Recipe No. 1: Cotton or Linen

Egyptian purples are claret-purples or brownish purples. This recipe is representative of those in which indigo is overdyed with madder or alizarin.

1. Scour material well, rinse, and leave wet.

2. Dye to a medium-dark blue with indigo (see indigo recipes, Blue chapter).

3. Mordant alum-tannin-alum with Basic Alum Mordant No. 2 or any cotton alum mordant.

4. Rinse well with fixing solution or water.

5. Dye with madder or alizarin as for Madder Red Recipe No. 1 (see Red chapter). Use madder at the rate of 1 pound madder or 8 to 10 grams of alizarin per pound of cotton. I generally use a mixture of madder and alizarin.

6. Remove, wash well, rinse, and dry.

• The madder red may be dyed first.

• The tannin may be omitted but only if aluminum acetate mordant is employed.

Egyptian Purple Recipe No. 2: Cotton or Linen

1. Scour material well, rinse, and leave wet.

2. Mordant with tannic acid at the rate of 1 to 1 1/2 ounces. The material should be in the tannin a minimum of 6 hours.

3. Remove from the tannin, squeeze well, and mordant with iron (6 to 8 teaspoons of ferric or ferrous sulfate in 4 to 6 gallons of room-temperature water). Mordant for 15 minutes, remove, squeeze, air, and mordant once or twice more.

4. Mordant the damp material with a weak basic alum. This can be previously used alum or new alum made with 4 ounces of alum per pound of cotton. Mordant for 6 to 12 hours.

5. Remove, squeeze well, and dry.

6. Treat with fixing solution or wash thoroughly.

7. Dye with madder (1 pound) or 10 grams of alizarin (see discussion of madder-alizarin dyebath preparation, Red chapter). Work the material in the room temperature or warm dyebath for about 30 minutes, then heat the dyebath slowly up to 160° to 170° F. Dye at the final temperature for about 1 hour.

8. Remove, squeeze out, wash with detergent, rinse well, and dry.

Egyptian Purple Recipe No. 3: Cotton or Linen

This is similar to the Pennsylvania German recipe above in that it uses tannin, alum, and iron as mordants, but in a simpler way. The recipe is my own.

1. Scour well, rinse, and leave the material damp.

2. Mordant with tannic acid at the rate of 1 ounce or equivalent

in sumac or other tannin source. Leave the material in the tannin a minimum of 6 hours.

3. Mordant with iron-contaminated alum. Buy the cheapest source of alum available. Usually, this will be aluminum sulfate purchased at a garden store, sold for acid-loving plants. If the compound does not appear to contain enough dark-colored material, add some copperas when making the solution. Too little iron will produce a claret-brown rather than a claret-purple. For best results, use the Basic Alum Mordant No. 2 formula at the rate of 12 ounces aluminum sulfate-copperas mix. The finished mordant should be light tan in color. Dilute the mordant enough to cover the material and mordant cold for at least 6 hours. Squeeze out thoroughly and hang the material to dry. (Clothespinning is the safest method with piece goods.) When thoroughly dry, mordant a second time, using the same lot of mordant.

4. Treat with fixing solution or wash thoroughly.

5. Dye with madder (1 pound) or 10 grams alizarin (see madder-alizarin dyebath preparation, Red chapter). Work the material in the room temperature or warm dyebath for about 30 minutes, then heat the dyebath slowly up to 160° to 170° F. Dye at the final temperature for about 1 hour.

6. Remove, squeeze out, wash with detergent, rinse well, and dry.

• Some shades of Egyptian purple are very nice for quilting, and go well with many other colors.

Madder Purple: Cotton or Linen

The color produced by this method is a reddish purple, and it is obtained by simply adding alkali, such as washing soda, to the dyebath until the color changes from orange or red to a bluish purple.

1. Scour well, rinse, and leave the material wet.

2. Select a madder or alizarin red recipe from the Red chapter, and premordant according to directions.

3. Prepare the madder or alizarin dyebath and add alkali until the bath is bluish purple.

4. Dye as for the madder red recipe and finish off as usual.

Cochineal Lilac and Pink-Purple: Cotton or Linen

Cochineal was seldom used on cotton, and then mostly by the calico printers. However, I believe that it produces some nice shades. The recipe given is the result of my own experimentation.

1. Scour well, rinse, and leave material damp.
2. Mordant with tannin, 1 ounce in 4 to 6 gallons of hot water. Mordant for 6 to 12 hours.
3. Remove, squeeze, and mordant with a basic alum for at least 12 hours.
4. Remove, squeeze, and dry.
5. Mordant with copper sulfate at the rate of 1/2 to 3/4 ounce. Add the copper sulfate to 4 to 6 gallons of hot water. Add the material and heat the bath to 170° to 190° F. Mordant at this temperature for 1 hour. Use a nonreactive, brass, or copper vessel.
6. Remove, squeeze, and dry or put the material directly into a cochineal dyebath prepared from 2 to 4 ounces of cochineal.
7. Work for a few minutes in the room-temperature dyebath, then heat the dyebath to about 170° F. Dye at this temperature for about 1 hour or until the color suits.
8. Remove, squeeze, rinse or wash, and dry.
 • Omission of the copper will produce a more crimson and less violet color.
 • This recipe may be used as an overdye with indigo.

Cochineal Crimson Recipe No. 1: Silk

Cochineal crimson, using alum mordant, has always turned out to be a very pleasing crimson-plum color. As it requires quite a bit of cochineal, the exhaust bath may be used for successive lighter batches, or for wool.

1. Scour silk or wet out well.
2. Mordant twice, with aluminum sulfate or aluminum acetate. Basic Alum Mordant No. 2 is also acceptable, as is potassium alum. Silk does not require ageing after drying.
3. Prepare a dyebath (4 to 6 gallons) using water and the dye-liquor from 6 ounces of cochineal.
4. Rinse the silk very thoroughly and add it to the room-temperature dyebath. Work well for a few minutes then heat the dyebath up to about 170° F. Dye at this temperature for about 1 hour.

5. If a brighter color is desired, dissolve about 1 teaspoon of tin (stannous chloride) in about 1 quart of water, lift and raise the silk, add the tin, stir, and reenter the silk. Addition of the tin may be done at any time, even before the silk is initially entered, or halfway through the dyeing procedure.

6. Remove, squeeze, wash with detergent, rinse well, and dry.

Cochineal Crimson Recipe No. 2: Silk

This recipe is essentially from Napier (also in *Dick's Encyclopedia*), and other early sources.

1. Scour material or wet out well and rinse.

2. Mordant with a tin (stannous chloride) solution. Dissolve 4 ounces of tin spirits or 6 to 8 grams of stannous chloride in soft or slightly acidified water (2 gallons). Pour the clear solution away from the sediment, if any, and warm the solution to about body temperature.

3. Add the silk. Work and mordant the material for about 1 hour. Remove and rinse well.

4. Prepare a cochineal dyebath, using 6 ounces of cochineal.

5. When the dyebath is at 100° F, add the damp silk. Work and dye the silk in the cochineal for 30 minutes, then sink the material under the dyebath liquor and leave overnight. If the shade is not blue enough to suit, add a little ammonia or cochineal dissolved in ammonia to the dyebath, reenter the goods, and dye for another 10 to 15 minutes.

6. Remove, wash out well in cold water, and dry.

• The same dyebath can probably be used for both Crimson Recipes 1 and 2, particularly if a little more cochineal is added before running Recipe 2.

• This recipe also works well if the dyebath is heated up to about 170° F and held at that temperature for 1 to 2 hours.

• Cochineal Crimson Recipes 1 and 2 yield nice additional purple shades when overdyed with indigo.

Cochineal Lilac: Silk

1. Scour or wet out well.

2. Mordant with tannic acid at the rate of 1 ounce in 4 to 6 gallons of water. Mordant for 6 to 12 hours.

3. Remove, squeeze, and mordant once or twice with aluminum sulfate or aluminum acetate. When dry, rinse very thoroughly.

4. Prepare a dyebath using 4 ounces of cochineal.

5. Add the silk to the warm dyebath, work well for a few minutes, then heat the dyebath slowly up to about 170° F. After about 30 minutes of dyeing at 170° F, dissolve a teaspoonful (6 grams) of tin (stannous chloride) in a quart of hot water, lift the silk, and add the tin. Stir well. Reenter the silk and dye for another 30 minutes. Remove, wash well, and dry.

• Alum mordants make cochineal silks more purple and less red; tin has the opposite effect. The shade of red-blue achieved is also affected by the pH of the dyebath. An acidic dyebath produces a redder color; addition of alkali, such as washing soda, changes the color of the dyebath to blue-purple. These color changes affect the shade of the dyed article as well.

Cochineal Red-Purple (Crimson): Wool

1. Scour the material well, rinse, and leave damp.

2. Prepare a dyebath using 1 to 2 ounces of well-ground cochineal.

3. Add the wool to the room temperature or lukewarm dyebath and work a few minutes.

4. Slowly heat the dyebath up to 190° to 200° F and dye at this temperature for at least 1 hour.

5. Remove, squeeze, rinse well, and dry.

• I believe that the prettiest crimson-lilacs on wool are produced this way (dyed substantively), and the prettiest cochineal purples result from overdyeing the material with indigo. Of course, if purple is the chosen end result, the preferable method would be to dye the material with indigo first.

• Premordanting with alum-tartar produces a redder and less blue shade than dyeing substantively.

• Cochineal, with or without indigo, produces fine purples on wool. Madder does not.

Cochineal Purple: Wool

Fine cochineal purples may be obtained by using a number of mordants and mordant combinations. For a thorough discussion of this subject see Gerber's book, *Cochineal and the Insect Dyes*. Premordanting with alum-chrome-tartar, followed by dyeing with 1 ounce of cochineal per pound of wool, produces an excellent result.

Cochineal-Indigo Dahlia: Wool

1. Scour wool well, rinse, and leave damp.

2. Premordant with alum-tartar at the rate of 2 1/2 ounces of alum and 3/4 ounce of tartar per pound of material. Mordant for 1 1/2 hours.

3. Remove from the mordant bath, squeeze out, and let the material hang on a line for 24 hours.

4. Prepare a 4 to 6 gallon dyebath, using the dyeliquor from 1 1/2 ounces of ground cochineal.

5. Heat the dyebath to about 170° F, add the wool, heat to 190° to 210° F, and dye at that temperature for 1 hour.

6. Remove, cool, and rinse.

7. Redye in a relatively weak indigo vat (hydrosulfite vat).

8. Redip until the color suits.

9. Soak wool in water for several hours, wash with soap or detergent, and dry.

• According to *Dick's Encyclopedia* (41), this recipe produces a fast purple much admired in German broadcloths.

• Dyeing first with the indigo, followed by mordanting and dyeing with the cochineal, gives a blue-purple.

8 Orange Dyes

Good early orange dyes usually involved an alum, alum-tin, or tin premordanted madder-yellow dye combination, the fastest yellow dye being weld (later, black oak bark). Kermes, lac, or cochineal were also used as the red dye. Good oranges on wool were also accomplished with tin-tartar mordant combinations with cochineal and black oak bark (Bancroft's Auroras). A number of yellow dyes produce off-oranges on chrome-mordanted wool. Fugitive oranges or yellow-oranges were gotten with annotto in South America and with saffron in Egypt, India, and the Far East. Iron buff can produce fast reddish orange, yellow-orange, and yellow shades on cotton and linen. Between 1810 and 1840 two extremely fast, but poisonous, mineral orange cellulosic dyes were produced. The first was antimony orange, developed by John Mercer (Floud, 1961), and the second, chrome orange. (John Mercer later developed a "dung fixing solution" substitute, composed of calcium and sodium phosphates, as well as the cotton "mercerization" process.) Antimony and chrome oranges were used on cotton until about 1910–20. Many old cotton quilts contain antimony orange, which possesses a very characteristic shade.

All recipes are for 1 pound of material.

Madder-Yellow Dye Oranges: Cotton or Linen

Many different shades of fast orange are possible if black oak bark is used for the yellow. Other more fugitive yellow dyes may also be substituted.

1. Scour material well, rinse, and premordant with aluminum acetate or Basic Alum Mordant No. 2. (If tannin is not used, aluminum acetate should be the alum mordant.) Mordant twice, allowing adequate ageing time.

2. Treat well with fixing solution or wet out thoroughly and rinse well several times.

3. Prepare a madder or alizarin dyebath (see Red chapter). Use a nonreactive, copper, or brass vessel.

4. Introduce the wetted-out material into the room-temperature dyebath. Work and dye the material for 30 minutes.

5. Lift out material and add a small quantity of yellow dye liquor (see Yellow chapter). Stir well and reenter the goods. (The yellow dye is added after first working the material in the red because yellow is taken up more rapidly than the red.)

6. Heat the dyebath slowly up to 160° to 180° F, as for madder red. Finish off as for madder red. If the color is too red to suit, more yellow dye may be added near the end of the dyeing procedure.

• Madder or alizarin alone produces an orange if the dyebath temperature remains between 130° and 150° F.

Madder or Madder-Yellow Dye Orange: Silk

The process is the same as for cotton-linen except that, with silk, you mordant once with strong alum, aluminum sulfate, or aluminum acetate—or twice with weaker mordant. Rinse well. Silk also takes a reasonably good orange with madder or alizarin alone, dyed at low temperature.

Madder-Yellow Dye Oranges: Wool

Madder produces very good oranges with a number of good, clear yellows. Suggested yellows are black oak bark, fustic, Osage orange, goldenrod, Queen Anne's lace, bidens, marigolds, and goldenrod.

1. Scour wool well and premordant with alum-tartar. The wool may be premordanted with tin-ox-tartar instead of alum, but the resulting orange will not be as lightfast.

2. Rinse well and dye with madder at the rate of 4 ounces of madder or 3 grams of alizarin.

3. Overdye with the yellow of your choice or lift the wool and add the yellow dye, cautiously, when the madder dyeing is completed or near completion. The yellow may also be dyed first, followed by the red.

4. Brighten with tin, if necessary, near the end of the dyeing procedure.

5. Remove, cool, rinse well, and dry. Wash with detergent if tin was used as an aftermordant.

• Premordanting with chrome often produces burnt orange. In this case, omit the tin afterbath.

- Remember that madder or alizarin alone, dyed at temperatures no higher than 140° F, gives an orange.
- A very good one-pot orange results by mixing black oak bark liquor, madder liquor, and "Bancroft's Mordant."

Iron Buff: Cotton and Linen

Iron buff (nankeen, nankin, rust, hydrated ferric oxide) has probably been in longer continuous use than any other fiber dye, with the possible exception of the tannins. It was used by the Swiss Lake Dwellers and by the ancient Egyptians. It was still in use by Japanese fiber artists in 1912 (Pellew, 1912, pp. 66–67) and in isolated villages in Germany and Poland up to 1940 (Brunello, Matthews). According to Pellew, the sails of many of the fishing boats on the Mediterranean around the turn of the twentieth century were dyed full shades with buff. It was also used by colonial Americans on their homespuns and in rugs and covertures by the French habitants on the St. Lawrence into the early part of the twentieth century. The dye is extremely fast to washing, exposure, and light. It can be yellow, reddish orange, or orange. It should be made in light shades only because the heavy shades are rougher to the touch and not entirely rubfast, and because large amounts of iron shorten fabric lifespan.

1. Scour material well, rinse, and leave damp.
2. Dissolve 2 level tablespoons of low-grade copperas per gallon of warm water. Three gallons of solution may be adequate for 1 pound of material. An iron or plastic bucket works well.
3. Work the material 15 to 30 minutes in the copperas bath, being sure to move the material frequently, particularly piece goods.
4. Squeeze out and work the material for about 15 minutes in a bath containing 3 teaspoons of washing soda per gallon of hot water (120° to 180° F). The iron will precipitate as a light greenish gray. Squeeze and air. In a few minutes the greenish gray color will change to orange or reddish orange-brown. Repeat the entire process if the color is not dark enough.
5. Wash well with detergent, rinse, and dry.

- Ferric salts (ferric chloride, nitrate, or sulfate) or "iron liquors" (nitrate or acetate of iron) may also be used. In these cases, the washing soda bath will probably not be needed. The "iron liquor" of the early Egyptians and colonial Americans was produced by steeping small pieces of scrap iron fillings in vinegar (ferrous acetate).

- Very pure copperas often gives a nice dull yellow (without the washing soda bath).
- It is possible to discharge iron buff completely, but great care must be exercised because the process requires strong acid (about a "1 normal" solution; pH less than 1). To do this, carefully pour 1 part of concentrated hydrochloric acid (muriatic acid) into 11 parts of water in a glass vessel and mix. This should be done outside or in a chemical fume hood. Next work the wet material with rods (wood or glass) for a few minutes in the acid. Then pick up the goods with rods and rinse 2 or 3 times at least. A final rinse with alkaline washing soda solution or diluted ammonia will help remove all traces of acid, which is mandatory. Otherwise, the fibers will be tendered.
- Do not dye cotton lint iron buff as it will be hard to card and spin.
- Historically, cottons and linens have been dyed iron buff by placing the material in iron water springs or burying the material in wet, iron containing mud—often for a week or two.
- Remember that iron buff dyed cotton is also iron mordanted, and can be dyed with a number of adjective dyes.

Annotto (Arnotto) Orange: Cotton, Linen, Wool, or Silk

Annotto is obtained from the orange-red outer coverings of the seeds of an indigenous South American shrub, *Bixa orellana*. The shrub seldom exceeds 12 feet in height. Around 1850, it was cultivated in Guiana, St. Domingo, and the East Indies. Actually, the shrub grows well in all tropical areas (Napier, 303–5, Adrosko, 28). The dye, Bixin, produces orange, yellowish-red, and salmon colors, but all are extremely fugitive in light and air, particularly on cotton and linen. Silks were dyed with the material at later dates (up to 1875–85) than cotton, linen, and wool. Annotto does fade true, if not subjected to acids or alkalis, but very rapidly. This dye has also been used to color butter and cheese—and our terrible-looking white oleomargarine during World War II.

1. Scour material well, rinse, and leave wet.
2. Add 4 to 6 gallons of soft water to a nonreactive, brass, or copper vessel. Add 1 to 2 ounces of annotto broken up fine (this may be put in a little fine mesh cotton bag).
3. Add 2 ounces (14 teaspoons) of pearlash or washing soda (for cotton, linen, or wool). For silk add 1/2 ounce of washing soda or pearlash, 1/4 ounce of natural white soap, and 2 ounces of annotto.
4. Boil or simmer the mixture for 1 hour. Then remove the bag

or allow the mixture to settle. If the annotto was free in the vessel, decant the clear.

5. Dye at scalding heat for cotton or wool, and at about 120° F for silk. Dyeing time will be 20 minutes to 1 hour. With cotton, it is sometimes advisable to dye for 20 minutes with two-thirds of the dyeliquor and then to add the last third of the dyeliquor and dye for an additional 20 minutes.

6. Remove, cool, rinse, and dry.

• Silk may require additional annotto for medium to dark shades.

• Addition of a small amount of acid or alum to the rinse water gives a reddish orange or salmon.

Chrome Orange: Cotton and Linen

Remember that this dye, like chrome yellow, is poisonous. Therefore, production of large quantities is not recommended.

1. Prepare a chrome yellow (see Yellow chapter). Squeeze and leave wet.

2. Prepare lime water or an alkaline solution of soda ash, about 2 teaspoons per gallon of water. Heat the alkaline solution to about 200° F. Add the wet chrome yellow dyed material. Work the material in the hot solution for about 10 minutes. The orange or orange-rose should develop almost instantly. Prolonged heating in a too-alkaline bath may discharge the color.

3. Remove, squeeze, wash well with detergent, rinse, and dry.

Antimony Orange: Cotton

To date, I have found only a single recipe for this dye (*A Practical Treatise on Dyeing and Calico Printing,* 136). Its production may have been done only in special factories, and I assume that the required chemicals are both unstable and dangerous. The final, stable product, is probably antimony sulfide.

Cochineal-Yellow Dye Oranges: Wool

Many yellow dyes, particularly those that incline towards the orange, such as late-season marigolds or old fustic, produce nice oranges when overdyed with cochineal. This usually works best by premordanting with alum-tartar, dyeing the yellow with tin afterbath, and then overdyeing with cochineal at the rate of 1 to 2 ounces of bugs per pound of wool. Turmeric dyed wool with tin afterbath, followed by cochineal, also produces a bright orange, but good lightfast qualities may be wanting. Thirty minutes' dyeing time each in the yellow and cochineal baths should be sufficient. Remember that, as a rule, cochineal requires a much higher temperature (simmer) than does the yellow dye.

Bancroft's Auroras: Wool

These oranges were first produced by Bancroft (1813) and studied by the Gerbers (Gerber and Gerber, 1974).

1. Scour wool well, rinse, and leave wet.
2. Premordant with tin-tartar or dissolve 1 1/2 ounces of cream of tartar in 1 gallon of soft water. Then add 3/8 ounce (11.5 grams, 2 teaspoons) of tin. Stir until dissolved. Add the tin-tartar to a nonreactive, brass, or copper vessel and add the dyeliquor from 1 ounce of black oak bark and 1 ounce of cochineal. Adjust the volume to 4 to 6 gallons. In the case of the premordanted yarn, prepare the dyebath from the black oak bark and cochineal liquors alone.
3. Add the wool and slowly heat the dyebath to the simmer. Dye at the simmer for about 30 minutes.
4. Remove, cool, wash with neutral detergent, rinse, and dry.
• By varying the amounts of bark and cochineal, quite a range of luminous oranges may be produced.

Orange with Black Oak Bark and Cochineal: Silk

This recipe is essentially from *Dyeplants and Dyeing* (1964).

1. Scour silk well or rinse and wet out thoroughly. Leave wet.
2. Prepare a dyebath of 4 to 6 gallons of soft water in a nonreactive, brass, or copper vessel.
3. Dissolve 1/2 ounce (14 grams, 4 teaspoons) oxalic acid in the dyebath. Then dissolve 1 ounce (28.5 grams, 5 teaspoons) tin. Finally, add the concentrated dyeliquor from 2 ounces of black oak bark. Stir well. Heat the dyebath to about 100° F.
4. Add the wet silk, work, and steep for 1 hour.
5. Remove the silk, squeeze, and air.
6. Add 1 additional ounce each of tin and oxalic acid to the dyebath. Stir until dissolved. Then add the concentrated dyeliquor from 2 ounces of cochineal. Heat the dyebath to 160° F.

7. Reenter the silk and steep at 160° F for an additional hour.
8. Remove, dry, rinse, and dry again.

Coreopsis or Bidens-Burnt Orange: Wool

This color is most distinctive and unique. It is one of our favorites. A recipe using coreopsis is listed in *Dyeplants and Dyeing*. Dale Liles has experimented with the recipe extensively and prefers the use of *Bidens polylepsis* (sunflower tickseed) over coreopsis.

1. Scour wool well, rinse, and premordant with chrome.

2. Prepare the dyebath by boiling 1 to 2 pecks of fresh or frozen coreopsis or tickseed flower heads in enough water to cover for 30 minutes. If *Bidens* is used, cut off the whole plant near the ground when the flowers are in bloom and prepare the dyebath by soaking at least 2 pecks (4 gallons) of packed plants overnight, then heat to the simmer for 60 minutes. Remove or strain the plant material.

3. Adjust the dyebath volume to 4 to 6 gallons.

4. Rinse the chrome-dyed wool, add the material to the dyebath, and slowly heat to the simmer. Dye until the burnt-orange color develops.

5. Remove, cool, wash well, rinse, and dry.

9 Brown Dyes

With dark browns, tans, fawns, and olives, a few of the traditional dyes were nearly as fast as the best of our modern dyes. Indian cutch, in my opinion, often surpasses the modern dyes in beauty on cotton and silk. This dye is obtained from two species of Indian Acacia trees, *Acacia catechu* and *Areca catechu*. A similar dye is obtained from Gambier, *Uncaria gambir* (Adrosko, 40). Cutch was used in Indian calico for centuries, and it was in use in European printed cottons by about 1800. The dye was still used in America by commercial dyers until about 1920. Apparently the trees also grow in places other than India, and according to Matthews (501), cutch was imported from Java, Singapore, the East Indies, and Peru. Cutch contains two dyes, catechu-tannic acid, $C_{38}H_{36}O_{16}H_2O$, and catechin, $C_{19}H_{20}O_2$ (Matthews, 501).

Another excellent natural brown dye on properly tannin-alum mordanted cotton and linen is American black walnut (*Juglans nigra*) or butternut (*Juglans cinera*) nut hulls. Black walnut or butternut hulls dye wool and silk beautifully with or without mordants. The dye in the green hulls—Juglone, 5-hydroxy-1,4-naphthoquinone, $C_{10}H_5O_2$ (OH)—is nearly colorless and oxidizes to the brown color. Use of black walnut hulls for dyeing in America probably goes back to the late 1600s both for fiber and as a hair dye.

Other natural brown dyes have been used throughout history. Fast natural brown was accomplished centuries ago in India by dyeing iron-alum mordanted cotton with Madder. By about 1820, manganese bronze (bistre), a mineral dye, came into use; it lasted until 1910 or slightly later. Shades from light bronze to full seal brown are possible, and the colors are fast. Logwood-fustic and madder-fustic combinations were also used for browns in the past, as were a number of barks.

All recipes are for 1 pound of material.

Cutch Brown Recipe No. 1: Cotton, Linen, or Silk

The following recipe is a modification of those found in Mairet, Adrosko, and several other sources.

1. Dissolve 4 ounces of cutch by boiling in 5 to 10 gallons of water, using a nonreactive, brass, copper, or iron vessel. This will take a little time because, while the catechu-tannic acid is soluble in cold water, the catechin is not. Therefore, the water must be quite hot to get all in solution. Make sure that no undissolved material is sticking to the bottom of the vessel prior to dyeing.

2. Dissolve 1/2 ounce of chrome (2 1/2 level teaspoons) in 4 to 6 gallons water in a nonreactive, brass, or copper vessel. Start heating the chrome up to 170° to 180° F.

3. When all the cutch is dissolved, remove the vessel from the heat and add 1/2 ounce (3 level teaspoons) of copper sulfate. Stir until dissolved.

4. When the dyebath temperature drops to about 170° F, add the well-scoured and wetted-out cotton or silk, or a combination of both. Work well for a few minutes, then sink the material. Apply heat in order to keep the dyebath between 160° and 180° F. If piece goods are in the dyebath, they will require more turning than yarn. Allow the items to remain in the dyebath for about 30 minutes.

5. Remove from the cutch, squeeze well, and place in the hot chrome. Work for a couple of minutes, then sink the material. You will notice the color becoming much darker as the chrome oxidizes the catechu and catechin to japonic acid. Of course, this chrome-mordants the material also, since the cotton or silk has absorbed tannin (cutch is about 40 percent tannin). Allow the material to remain in the chrome for about 10 minutes.

6. Remove from the chrome, squeeze well, judge the shade of the goods and evenness of dyeing. Chances are that the shade will be too light and, if you are working with piece goods, the dye job may not be completely level. If either or both are the case, place the material back in the cutch bath for another 20 to 30 minutes. Work well if the dye job is uneven.

7. Remove, squeeze, and place in the chrome bath. Repeat the process again if the shade is still not dark enough.

8. Remove, squeeze, wash well with detergent, rinse, and dry.

- Silk dyes to the same shade in this dye in less time than it takes for cotton or linen. Therefore, work well to ensure level dyeing or use a weaker dyebath—say 2 to 3 ounces of cutch.

• The same dyebath may be used for 1 or 2 pounds more of material, but only for lighter shades. If a series of shades is desired, the exhaust baths work extremely well. I usually add a little more copper with each exhaust bath.

• As is true with all natural fibers, one type of cotton or silk will dye to deeper shades than another. By using this method, you can more likely get the desired shade. If not, see Recipe No. 2.

Cutch Brown Recipe No. 2 (Cutch-Yellow Dye, Cutch-Logwood, or Cutch-Logwood-Yellow Dye): Cotton, Linen, or Silk

With certain yarns or piece goods, cutch alone may not answer, and it is easier to get the desired effect with combination. Combination also frequently requires less cutch. One good way is to use Recipe No. 1 but have on hand dyeliquor from 1 ounce of logwood, along with some good yellow dye. If the shade or depth of brown is not great enough, add some yellow dye. This will produce a darker brown. Logwood liquor will darken the product even more. Quite dark shades result from use of cutch, logwood, and a yellow dye. Recipe No. 2 is somewhat similar to that of Mairet (50).

1. Prepare a cutch bath as in Recipe No. 1, but with only 2 ounces of cutch. Do not add copper sulfate at this point.

2. Prepare a chrome bath as in Recipe No. 1.

3. Work the material in the cutch bath for 20 to 30 minutes, then in the chrome for 10 minutes.

4. Add the dyeliquor from 3 ounces of fustic chips or 1 ounce of fustic sawdust. Stir. A little logwood liquor may also be added.

5. Reenter the cotton in the cutch-fustic bath and dye for another 20 minutes.

6. Remove, squeeze, and work in the chrome bath.

7. Add 1/2 ounce of copper sulfate to the cutch-fustic bath and stir until dissolved.

8. Remove from the chrome bath, squeeze well, and place the material in the cutch-fustic. Dye for at least 20 to 30 minutes more.

9. If the material is dark enough and even, remove, squeeze, wash well with detergent and dry. If not dark enough, add logwood liquor and dye some more.

• Addition of a little copperas to the dyebath will have a greying effect.

- This recipe, and the preceding one, can be done without use of chrome, but the color will not be as dark or fast. If chrome is omitted, make certain that copper is added to the dyebath.
- Wool dyes very well using 1 1/2 to 3 ounces of cutch. Add copper sulfate to the dyebath.

Black Walnut: Cotton or Linen

1. Collect the nuts of either black walnut or butternut when the nut husks are still green. It is best to pick up the nuts soon after they have fallen from the trees. Put on rubber gloves and remove the rinds with a knife. The green rinds can be used for the dyebath immediately or spread out on newspaper, dried, and stored for later use. If the dyebath is to be prepared immediately, the green nuts, husk and all, may be used. I have also stored the whole green nuts or rinds for extended periods in a deep freeze.

2. Premordant cotton or linen with alum-tannin-alum. Very inexpensive alum or aluminum sulfate works very well since some iron contamination is beneficial with this dye.

3. Add 2 gallons of green rinds, 1 1/2 gallons of dry rinds, or 4 to 5 gallons of nuts to a vessel capable of holding the material, and one which may be heated. An iron pot is fine for walnut. Cover with water, add a little vinegar, acetic acid, or vitamin C (ascorbic acid) and soak overnight. Heat the vessel to the simmer or boil the next day and keep at that temperature for 2 hours. Cool, strain out the rinds or nuts, and bring the liquid volume up to 4 to 6 gallons.

4. Rinse the mordanted goods thoroughly or use fixing solution first and then rinse. Squeeze out.

5. Add the cotton or linen to the dyebath. Initial dyebath temperature may be room temperature to about 140° F.

6. Work the material well and heat the dyebath up to 170° to 212° F. Dyeing time will probably be 1 to 2 hours.

7. Remove, squeeze, wash with detergent, rinse well, and dry.

- All parts of the walnut or butternut tree contain the dye; the highest concentration is in the nut hulls. A dyebath may be prepared from leaves. If so, use a lot of them, as well as a pinch of copperas and a tannin source.
- Rinds of nuts from different trees vary as to dye content, and some produce a yellow-brown rather than dark brown. If a yellow-brown results, and a dark brown is desired, add a little tannin or tan-

nin source such as sumac leaves or berries and a pinch of copperas, and dye for an additional length of time.

• Black walnut is an excellent dye for wool and silk. Often these dye substantively to beautiful shades. Alum-premordanted wool gives a yellowish brown, and chrome a darker greenish brown. Directions for brown on wool with walnut by Viner and Viner (59), an old Appalachian mountain recipe, are as follows: "For a pale tan boil for a few minutes only; for darker shades continue boiling for from 10 minutes to an hour or more. To get a very dark brown add a pinch of copperas and a double handful of sumac berries. Simmer the wool in this for an hour or more and leave in the ooze overnight. Then rinse very thoroughly, both the wool and dyepot before it is used for other work." Do not add too much copperas or a gray will result.

• A good black may be obtained by overdyeing a deep indigo blue with strong walnut. In this case sumac leaves or berries or a little tannin, and a little copperas, are added to the walnut. This works well on cotton, wool, or silk.

Madder (Alizarin) Brown: Cotton and Linen

All madder or alizarin colors are quite lightfast, but it may be a bit tricky getting brown. Browns, grays, and violets are obtained with madder using a combination of alum and iron mordants. The following recipe is suggested, at least as a starting point for experimentation.

1. Make up a quantity of Basic Alum Mordant No. 2, using the most contaminated aluminum sulfate you can find. Such alum is sold in garden stores for acid-loving plants. If it doesn't look as if it contains enough dark particles (iron contaminants), add some copperas. The product will mordant the material with both aluminum and iron. Make up the mordant at the rate of 8 to 12 ounces of aluminum sulfate per pound of cotton to be dyed.

2. Mordant alum-tannin-alum, using the same batch of alum mordant.

3. Treat with fixing solution or wash very thoroughly.

4. Prepare a madder bath at the rate of 8 ounces of madder per pound of cotton (see madder red recipes) or use at least 4 ounces of madder and 2 to 3 grams of alizarin or use 5 to 6 grams of alizarin.

5. Add the damp cotton and heat up to the boil, or at least to a strong simmer. Keep at this high temperature for about 1 hour or more. In this case the idea is to extract the brown xanthins in the madder. A fast brown should develop. If not, you may have a nice Egyptian purple.

6. Remove, squeeze, wash and rinse well, and dry.

• Addition of yellow dye late in the dyeing procedure helps achieve the brown.

Manganese Bronze, "Bistre," "Gunstock": Cotton, Linen, Silk, or Wool

Manganese brown was used extensively on cotton, but it also works well on wool, skins, and fur. It is extremely fast to light and washing, but can be discharged with Rit or other color remover (alkali and sodium hydrosulfite or thiourea dioxide).

Pellew (1912, p. 96) considers the dye especially useful to crafts-persons, not only for full shades of brown but also for light tan (honey) shades, as a cover, or to soften and blend harsh combinations of other colors. Pellew also describes a unique application for manganese bronze that was used during the Boer War (1899–1902). At one point, it was necessary for the English to send as many mounted troops to the front as possible, including the famous Scots Greys, who rode on white or light gray horses. Someone in the British war department thought that white horses would be too good a target for the Boer marksmen, so a dye chemist was consulted. The chemist advised that several kegs of potassium permanganate (purple crystals) be sent down to the troopships. Each trooper was instructed to sponge his horse with a weak solution of the chemical every morning. Apparently it worked because the hair and skin of the horses was a nice soft shade of brown long before the troopships reached South Africa.

Manganese brown is the easiest of all dyes to produce, in my experience, if done by the purple crystals method developed by Pellew.

1. Scour the cotton or wool, rinse, and leave damp.

2. Dissolve 1 to 2 tablespoons of purple crystals in 1 to 2 quarts of hot water. When all is dissolved, pour the fluid into 4 to 6 gallons of room-temperature water in a plastic vessel. Stir.

3. Put 4 to 6 gallons of water in another plastic vessel, and add 1 to 2 ounces of Karo or other sugar syrup or 3 to 4 teaspoons of glucose. Either syrups or molasses may be used. Both contain reducing sugars that stop the action of the permanganate, which is an oxidizing agent. The oxidizing action of the purple crystals needs to be stopped or the material could become weakened.

4. Add the scoured, wet yarn or piece goods to the cold permanganate bath and work for about 5 minutes, squeeze out well, and

plunge the material directly into the syrup bath. Work for about 1 minute.

5. Remove from the syrup, squeeze, and check for depth of color and levelness of dyeing. If the color is not deep enough, or the dyeing uneven, repeat the process. In fact, the process may be repeated as many times as necessary. If the color does not get dark enough after repeated dips, add more purple crystals.

6. Finish off in the syrup bath, then wash with detergent, rinse well, and dry.

• This recipe produces darker shades on wool than on cotton or linen. Much darker shades result on cotton or linen which has been premordanted with tannin at the rate of 1 to 1 1/2 ounces per pound of cotton. I have not seen this last process written up anywhere; it turned up quite by accident at a traditional cotton dye workshop I was teaching. I have labeled the darker dye "gunstock" because one of the traditional gunstock stains was produced by brushing purple crystals solution directly on the bare wood. Of course, the wood contained the tannins. I recommend "gunstock" highly for dark shades of brown.

Bark Brown Recipe No. 1: Cotton, Linen, or Wool

A number of barks were used by our ancestors for brown on cotton, linen and wool. Barks used included maple, several of the red and white oaks, hemlock, and alder. Typically, about 1 peck (2 gallons) of bark was soaked overnight and then boiled for about 2 hours. Then the bark was removed and the volume adjusted to 4 to 6 gallons. Into this was placed 1 pound of alum-tartar or chrome premordanted wool or alum-tannin-alum premordanted cotton. The dyebath was then heated to the simmer or boil and kept at that temperature for 30 to 60 minutes. Some recipes (Adrosko, 72) called for boiling the dyed item, without rinsing, in a weak afterbath for about 10 minutes. The afterbath contained 1/6 ounce of chrome, copper sulfate, or copperas. In the case of chrome or copper, 6 to 7 tablespoonfuls of vinegar were also added. Such use of afterbaths probably renders the dyed goods more light- and washfast.

Bark Brown Recipe No. 2: Cotton or Linen

1. Scour material well with soda ash (sodium carbonate), remove, wring out, but do not rinse.

2. Prepare the dyebath by boiling 1 1/2 gallons of white oak or maple bark for two hours in 4 to 6 gallons of water. Any vessel capable of being heated is satisfactory.

3. Remove the bark and strain the dyeliquor through cheesecloth.

4. When the dyeliquor temperature has dropped to 140° to 150° F, add 3 ounces of copperas and stir well.

5. Add the damp cotton and work and dye the material for about 15 minutes (do not heat the dyebath). Then remove, squeeze, and air for about 5 minutes.

6. Repeat (step 5) 3 additional times.

7. Remove, squeeze, wash well with detergent, rinse well, and dry.

Mineral Khaki: Cotton or Linen

The name *khaki* is derived from an East Indian term meaning "dirt" (Matthews, 515). The dye was used mostly for army cloth and clothing because it made them difficult to see at a distance. The dye itself, a mixture of iron and chromium oxides, is very light- and washfast. It is also resistant to alkalis but not to strong acids, which discharge the iron and chromium oxides.

Mineral khaki was abandoned for most army uniform cloth just prior to World War I. The reason was that the new synthetic sulfur colors, which were intensively researched between 1890 and 1910, produced good, fast khaki shades, and the cloth was more comfortable to the wearer (all mineral dyes roughen cloth, at least slightly).

1. Prepare a room-temperature dyebath (4 to 6 gallons) by dissolving 1 ounce of ferric nitrate, sulfate, or chloride. Then add 2 ounces of chrome alum. One tablespoon of ferric sulfate weighs about 1/3 ounce, and 1 tablespoon of chrome alum weighs about 1/2 ounce. Stir until the materials are completely dissolved.

2. Work the wet, well-scoured cotton or linen in the dyebath for about 30 minutes.

3. Remove, squeeze out, and place the material in a vessel capable of being heated, containing 4 to 6 gallons of washing soda liquor. To prepare the washing soda liquor, dissolve 1 1/2 ounces of washing soda in the 4 to 6 gallons of water.

4. Heat the material in the washing soda liquor up to the boil.

5. Remove, squeeze, wash well with detergent, rinse, and dry.

• Most shades of khaki can be produced by varying the relative amounts of the chromium and iron salts, or by the addition of a small amount of manganese chloride or sulfate to the first bath.

Madder Brown or Off-Brown: Wool

A reddish-brown is obtained by premordanting with chrome. Addition of 2 ml concentrated sulfuric acid or 1 1/2 ounces tartaric acid to the mordant bath will make the color darker. For 1 pound of wool use about 8 ounces of madder or 6 grams of alizarin. This dye may also be done by the one-pot method, but the color will be somewhat yellower and not as deep (Hummel, 345). Remember to add one teaspoon of chalk to the dyebath unless the water is very hard. Dye at 180° to 200° F for about 2 hours. Premordanting with copper sulfate-tartar gives brown and copperas-tartar yet darker browns. These dyes may work better with madder than alizarin.

Combination Brown Recipe No. 1: Wool

In former times some of the nicest browns were produced using a combination of dyes and mordants. These had names both descriptive and charming, such as "Paris mud," "London smoke," "liver," "bat wing," and "snuff brown." The following recipes are representative. (Recipes are given in Bemis and Molony.)

1. Prepare a 4 to 6 gallon dyebath in a nonreactive, brass, copper, or iron vessel. Add the dyeliquor from 1 1/2 ounces of madder or 1 gram of alizarin and 4 to 5 ounces of old fustic chips or 3 ounces of sawdust. Add 17 grams (4 teaspoons) of alum and 9 grams (3 teaspoons) of cream of tartar. Boil this mixture for about 15 minutes, remove from the fire, and allow to cool to about 150° to 160° F.

2. Add the well-scoured, damp wool, and heat up to the simmer. Dye at the simmer for 1 hour.

3. Lift the material and add 6 grams (1 1/2 teaspoons) of copper sulfate and 3 grams (1 teaspoon) of copperas. Stir until dissolved, then reenter the goods and dye for 30 minutes longer.

4. Remove, cool, wash well, and dry.

Combination Brown Recipe No. 2: Wool

1. Prepare a mordant bath by dissolving 6 grams (1 teaspoon) of chrome, 6 grams (2 teaspoons) of cream of tartar and 6 grams (1 teaspoon) of alum in 4 to 6 gallons of water in a nonreactive vessel. Mordant at the simmer for 1 to 1 1/2 hours.

2. Remove, cool, and rinse very well.

3. Prepare the dyebath by adding the dyeliquor from 3 ounces of fustic, 1 1/2 ounces of madder or 1 gram of alizarin, 3/4 ounce of brazilwood, and 1/2 ounce of logwood. Add enough water to bring the dyebath volume up to 4 to 6 gallons. Add the mordanted goods and heat the dyebath up to the simmer. Dye at the simmer for about 40 minutes.

4. Remove, cool, wash well, and dry.

• A number of shades may be produced, particularly by varying the amount of fustic and logwood. These browns are very nice.

Combination Brown Recipe No. 3: Wool

1. Prepare a dyebath by adding the dyeliquor from 3 ounces of fustic, 3 ounces of madder or 2 grams of alizarin, 1 1/2 ounces of brazilwood, and .4 ounce of logwood to enough water to make 4 to 6 gallons. The dye vessel may be nonreactive or iron.

2. Add the well-scoured, damp wool and heat the dyebath up to the simmer. Keep the dyebath at the simmer for about 40 minutes.

3. Lift the material and add a solution of copperas. Prepare the copperas by dissolving 6 grams (2 teaspoons) of copperas in a quart of hot water. Stir well.

4. Reenter the goods and dye for 30 minutes more.

5. Remove, cool, wash well, rinse, and dry.

10 Gray and Black Dyes

The grays and blacks of antiquity were practically all iron-tannates, fibers treated with iron salts and vegetable tannins as from sumac, hemlock, or galls. Properly done, these grays and blacks often lasted for years, though they could revert to a rusty color if subjected to excessive exposure and sunlight. For grays on cellulosics and silk, iron-tannates often continued to be used until nearly the turn of the twentieth century. By 1600, logwood from Central and South America was often added along with iron tannate and sometimes copper salts, producing a blacker and sometimes more permanent color. Good early blacks were also produced by combining the three primary colors, using indigo, madder, and a good yellow dye. Good blacks also resulted by dyeing strong black walnut or cutch on a deep indigo bottom. When logwood became available it was much used for gray and black, often in combination with old fustic or old fustic and madder. Indeed, logwood, properly mordanted, was used for inexpensive blacks until about 1940, longer than any other natural dye (Adrosko, 47). Matthews devotes an entire chapter to the use of logwood.

Good grays and blacks have often caused natural dyers trouble. Therefore, I have researched and experimented with them rather extensively. The good shades are very beautiful.

All recipes are for 1 pound of material.

Black Walnut or Cutch on Deep Indigo Bottom: Cotton, Linen, Silk, or Wool

1. Scour material well, rinse, and leave wet.
2. For silk, cotton, or linen, dye and redip in the zinc-lime indigo vat (see Blue chapter) until the material appears to be "black as coal." Dye wool in the hydrosulfate or thiourea dioxide vat until very deep blue-purple.
3. For linen, cotton, and silk, mordant with tannin at the rate of

1 ounce of tannin per pound. Then mordant with iron (see general mordanting procedures). For wool, mordant with iron-oxalic acid unless a slightly lighter purple-black is desired.

4. Dye with a strong black walnut dyebath (see Brown chapter). Use somewhat lower dyeing temperatures (160° F) for cotton, linen, and silk than for wool (170° to 190° F).

5. Remove, cool, wash well with detergent, and dry.

• With cotton, linen, and silk, cutch may be used instead of black walnut (see Brown chapter). Strong cutch dyebaths make wool feel harsh and are not recommended. If cutch is used, omit the mordanting step with tannin since cutch is about 40 percent tannin.

Combination Deep Black: Cotton or Linen

This recipe is modified from Mairet (51). By adding more madder, a series of reddish blacks may be produced. Part of the function of the red is to offset the greenish tinge which the logwood might develop in time.

1. Scour material well, rinse, and leave wet.

2. Mordant 10 to 12 hours with tannin, 1 ounce per pound of cotton.

3. Wring out, and work for 10 minutes in a washing soda bath at 120° to 140° F. To prepare the bath dissolve 1 to 1 1/2 ounces of washing soda in 4 to 6 gallons of hot water. Do not discard the soda bath.

4. Wring out, and work for 30 minutes in a copperas bath of 1 1/4 ounces copperas dissolved in 4 to 6 gallons of hot water.

5. Wring out and return the material to the washing soda bath for 15 minutes.

6. Wash well and leave wet. At this point the material should be a rather nice dark gray.

7. Prepare a dyebath using the dyeliquor from 12 ounces of logwood chips or 6 ounces of sawdust, 2 1/2 ounces of madder or 1 gram of alizarin, and 8 ounces of fustic chips or 4 ounces of fine fustic sawdust or 1 ounce of fustic or Osage orange extract. Final dyebath volume should be 4 to 6 gallons.

8. Enter the goods into the 80° to 110° F dyebath. Slowly raise the temperature to the simmer or boil. Keep the temperature at the simmer or boil for 30 minutes.

9. For a blacker and more lightfast dye, work the wet, dyed goods for 5 to 10 minutes in a hot (160° to 180° F) chrome or copper bath. Make the bath by dissolving 1/2 ounce of chrome or copper in 4 to 6 gallons of hot water.

10. Remove, squeeze, wash well with detergent, rinse well, and dry.

• The dyebath and mordant baths may be used a second time for a lighter shade.

• Addition of a little "black iron liquor" along with the copperas produces a richer color.

Jet Black: Cotton or Linen

1. Scour material well, rinse, and leave wet.

2. Mordant with 1 ounce of tannin or 4 to 5 ounces of sumac or equivalent. Leave in the mordant overnight.

3. Wring out and work for 10 minutes in lime water or hot (120° to 140° F) washing soda bath (1 ounce of washing soda in 4 to 6 gallons of hot water). Do not discard the alkaline bath.

4. Wring out and work for 30 minutes in a hot (120° to 140° F) mordant bath produced by dissolving 1 1/2 ounces of copperas and 3/4 ounce of copper sulfate in 4 to 6 gallons of water. The copper sulfate improves lightfastness. If alum is used instead of or in addition to copper, the black will have a slight purplish cast.

5. Wring out and work for an additional 10 minutes in the alkaline bath. At this point the material should be dark gray.

6. Wash the material well and leave wet.

7. Prepare a dyebath using the dyeliquor from 5 ounces of logwood chips or 3 ounces of sawdust and 1 1/2 ounces of old fustic or Osage orange chips or 3/4 ounce of sawdust. For best results add 1 gram of alizarin or 1 to 2 ounces of madder as well. Adjust the volume to 4 to 6 gallons, then add 1 teaspoon of washing soda and stir well.

8. Enter the goods and heat the dyebath up to 140° to 180° F. Dye at this temperature for 30 minutes.

9. Lift the goods. Add 6 grams (1 1/2 teaspoons) of copperas dissolved in 1 quart hot water. Stir, reenter the goods and dye for an additional 10 minutes.

10. Remove, squeeze out, and for the most washfast and clearest black, work the goods for 10 to 15 minutes in a hot (180° F) chrome mordant bath, about 7 grams (1 1/2 teaspoons) of chrome in 4 to 6 gallons hot water.

11. Remove, squeeze out well, wash with detergent, rinse well, and dry.

• Cutch dyeliquor may be substituted for the yellow dye. Use about 2 ounces.

- Omission of the yellow or brown dye will produce a black, but not as intense. The same may be true if the alkaline baths are omitted.
- Weaker dyebaths will give grays and gray-blacks.
- This recipe gives an intense, deep black that will not fade. However, if the yarn or piece goods disintegrates in 50 to 100 years because of "iron tendering," don't say that you haven't been warned! This effect often appears in black cottons and silks in antique quilts, embroidered fabrics, etc.

Logwood Black, One Pot: Cotton or Linen

This recipe is included *but not recommended.* The method is discussed in Matthews (489) and is probably typical of industrial methods used during World War I. Note: *The directions are for a test skein of cotton only, not one pound of material.*

1. Dissolve 1.5 grams of chrome in 1 ounce of water.
2. Mix this with 1 pint of rather strong logwood dyeliquor (.1 ounce of logwood sawdust). Then add 3.5 ml of hydrochloric acid.
3. Add a small skein of well-scoured, rinsed, and wet cotton.
4. Heat the dyebath slowly up to the boil. The dyeing time should be about 1 hour at which time the cotton should be deep blue.
5. Next work the skein in a calcium acetate bath (about 1 gram in 1 pint of water) for about 30 minutes. The color should change to a blue-black.

- Chrome blacks with logwood are very fast to washing, but not too fast to light, usually developing a greenish cast in time, particularly if no madder or alizarin is added to the dyebath. Recently, Mary Frances Davidson, author of *The Dyepot,* gave me some of her old dyestuffs. One was a bottle of haematoxylin (logwood) with the following comment: "Logwood Blacks fade to gray or green. I know— my stockings in W. W. I were green in time."

Tannin-Iron Black: Cotton or Linen

This basic recipe is the earliest black of the ancients, used for both cotton and silk. It is listed primarily for historical purposes and is not necessarily recommended since the amount of iron necessary to give a deep black will markedly decrease the lifespan of the yarn or fabric. Such a deep black on silk will weight the material markedly. The shades of black are also not particularly attractive.

1. Scour material well, rinse, and leave wet.

2. Prepare a tannin bath, using technical tannic acid (1 to 2 ounces) or equivalent of cutch, sumac, myrobalans, etc.

3. Mordant the material in the tannin for about 6 hours. Do not throw out the tannin.

4. Squeeze out, rinse, and mordant for 30 minutes in an iron bath, about 4 ounces of copperas for silk or 2 ounces for cotton in 4 to 6 gallons of room-temperature water. Ferric sulfate, or nitrate or acetate of iron, or ferric chloride may be substituted for the copperas.

5. Squeeze out, rinse, and place the material back in the tannin bath for 1 hour.

6. Repeat in the iron solution.

7. Go back and forth between the tannin and iron until the desired depth of black is reached. Additional tannin and iron may need to be added to the respective baths.

8. Wash well with detergent, rinse well, and dry.

• Nice shades of gray often result during the early stages of this recipe.

• With cotton, the material may be worked for 10 minutes in a lime water or washing soda bath following treatment with the tannin. This helps fix both the tannin and iron and hastens the darkening process. The process is not recommended for silk as it may destroy all luster.

Logwood-Cutch Black: Silk

By 1918–20, logwood was still extensively used to dye silk, whether weighted or unweighted. These blacks utilized tannin-iron mordant, the tannin source often being cutch. Logwood renders the silk fiber opaque, thus producing a full and brilliant black, whereas the coal-tar blacks of the period produced a dull faded appearance (Matthews, 481). Linen also has a tremendous affinity for logwood. On silk, most natural colors other than blacks had been replaced by synthetic dyes by 1920, except where considerable weighting (with metallic mordants) was desired. Unfortunately, heavy mordant weightings are often injurious to the luster and softness of silk fibers and shorten their lifespan.

1. Scour or wet out silk well, rinse, and leave wet.

2. Dye the silk dark cutch brown (see Brown chapter). Use 3 ounces of cutch.

3. Squeeze out and mordant with copperas, ferric sulfate, or black iron liquor. Dissolve about 2 ounces of copperas in 4 to 6 gallons of

warm water. Work the material in the mordant for approximately 20 to 30 minutes.

4. Prepare a logwood dyeliquor by boiling 8 ounces of logwood chips or 4 ounces of sawdust for about 45 minutes. Add the dyeliquor to 4 to 6 gallons of room-temperature water.

5. Add the silk and slowly raise the temperature to about 160° F. Dye for 30 minutes at 160° F.

6. Remove, squeeze out, wash well with neutral detergent, rinse well, and dry.

Logwood-Fustic Black: Silk

By 1918 the number of logwood silk gray and black recipes was legion. This recipe is quite typical of those used from about 1850–1920. Osage orange may be substituted for the fustic. If the yellow dye is omitted, a lighter black or dark gray results. Omission of the yellow dye and less logwood produces a gray.

1. Scour well, rinse, and leave wet.

2. Mordant well with copperas, about 3 ounces in 4 to 6 gallons of warm water. Add a little (1/2 ounce) nitrate of iron if available. Mordant for 1 hour.

3. Remove, squeeze out, rinse well.

4. Prepare a dyebath using the dyeliquor from 1 pound of logwood chips or 8 ounces of sawdust and 3 ounces of fustic chips or 1 1/2 ounces of sawdust. Heat the dyebath to no more than 160° F. Dyeing time should be about 1 hour.

5. Lift, and add 1/2 ounce of copperas (previously dissolved), re-enter the goods and dye for an additional 10 minutes.

6. Remove, squeeze out, wash well with neutral detergent, rinse well, and dry.

Blue-Black: Silk

A bluish shade of deep black may be obtained by first dyeing with indigo or Prussian blue and then dyeing with logwood. On silk, Prussian blue generally possesses the better sheen and luster, probably because the dyevat is acidic rather than alkaline as with indigo. This is my favorite silk black.

1. Dye silk a deep Prussian blue (see Blue chapter).

2. Mordant for 30 minutes with copperas, 1 1/2 ounces in 4 to 6 gallons of room-temperature water.

3. Remove, squeeze out, and rinse well in cold water.

4. Prepare a dyebath using the dyeliquor from 6 1/2 ounces of logwood chips or 3 1/2 ounces of sawdust. Adjust the volume to 4 to 6 gallons.

5. Add the silk and heat the dyebath to no higher than 160° F. Dye for about 30 minutes.

6. Lift the material and add about 1 quart of the copperas mordant. Stir, reenter the goods and dye for an additional 10 minutes.

7. Remove, squeeze out, wash gently with neutral detergent, rinse well, and dry.

Combination Black: Silk (Unweighted)

All of the preceding silk black recipes will weight the material 30 to 100 percent or more. The following recipe, known as "pure dye" black, adapted from Matthews (482), should produce a lesser degree of weighting.

1. Scour or wet out silk thoroughly, rinse, and leave wet.

2. Prepare a dyebath using the dyeliquor from 12 ounces of logwood chips or 6 ounces of sawdust, and 3 ounces of fustic chips or 1 1/2 ounces of sawdust. Add 3/4 ounce of copperas and 1/2 ounce of verdigris or copper sulfate. Add sufficient water to make 4 to 6 total gallons. Heat the dyebath to 160° F.

3. Add the wetted out silk and dye for at least 30 minutes.

4. Remove, squeeze out, and hang to "age" in the air for 1 or more hours.

5. Prepare a new logwood dyebath using 12 ounces of logwood chips or 6 ounces of sawdust. Adjust the volume to 4 to 6 gallons. Add 2 ounces of natural soap (not detergent), finely shaved. Heat the dyebath to 160° F.

6. Add the silk and dye for 1 hour.

7. Remove, squeeze, rinse well, and dry.

Black from Primary Colors: Cotton, Linen, Silk or Wool

1. Scour well, rinse well, and leave material wet.

2. With cotton, linen and silk, dye a "coal black" indigo blue-purple, using a strong zinc-lime indigo vat. Use several dips. With wool, use the hydrosulfite indigo vat. Dye until the wool is mazareen blue or darker (see Blue chapter). Wash well.

3. Mordant with alum-tannin-alum (cotton, linen, silk) or alum-tartar (wool).

4. Prepare a dyebath of yellow dye (fustic, Osage orange, goldenrod, black oak bark) and madder or alizarin. Two ounces of madder or 1 gram of alizarin and 1/2 to 1 ounce of black oak bark or 4 ounces of fustic or Osage orange chips or 2 ounces of sawdust should suffice. Adjust the dyebath volume to 4 to 6 gallons.

5. Enter the goods into the room-temperature bath and slowly heat up to the simmer (wool, cotton, linen) or to 160° F (silk). Dye the material at the high temperature for 30 to 60 minutes.

6. Remove, squeeze, wash well with neutral detergent, rinse well, and dry.

• This recipe and the following one do not use iron mordants, which at times are a bit hard to apply evenly and can have a rather corrosive effect on fibers.

Black from Primary Colors: Cotton or Linen

It is reasonable to assume that a good black should be obtained by combining logwood blue with madder and a yellow dye. Such a recipe exists in Hellot/Macquer (483). The recipe calls for weld, the fastest yellow dye of antiquity; other yellow dyes may be substituted, but black oak bark and weld are the best.

1. Scour well, rinse, and leave wet.

2. Mordant with alum-tannin-alum.

3. Prepare a yellow dyebath to give a medium shade of yellow, about 1/2 ounce of black oak bark, etc. (see Yellow chapter). Dye the yellow at 120° to 140° F.

4. Squeeze and prepare a logwood dyebath. Use sufficient logwood to produce a dark blue (8 ounces of chips or 4 to 5 ounces of sawdust). Boil or simmer the logwood for about 45 minutes. Add sufficient water to produce a dyebath of 4 to 6 gallons.

5. Add 1 to 1 1/2 ounces of copper sulfate, previously dissolved in 1 quart of hot water. Stir.

6. Add the wet goods, heat the dyebath and dye at 160° to 190° F for about 45 minutes, turning the material frequently.

7. Remove, squeeze, and rinse well.

8. Now place the material in a dyebath produced from 3 to 4 ounces of madder or 1 to 2 grams of alizarin. Place the material in the room-temperature dyebath and slowly heat up to 160° to 180° F. Dye at this temperature for 1 hour.

9. Remove, squeeze, wash well with detergent, rinse well, and dry.

• The madder liquor and logwood liquor may be used in the same dyebath, if the material is premordanted with copper following the yellow dyeing.

Common Black: Wool

This is a very common old wool black. It should be fairly fast to light because of the use of both iron and copper mordants.

1. Scour well, rinse, and leave wool wet.

2. Bring 4 to 6 gallons of water to the boil or simmer and add 2 ounces of copperas and 3.5 grams (1 teaspoon) of copper sulfate. Stir until dissolved.

3. Add the yarn or cloth and work frequently for 1 hour. Keep the temperature at the simmer or boil. Raise the material and air well once, halfway through the mordanting.

4. Remove, drain, squeeze out, air well, and rinse thoroughly.

5. Discard the mordant. Add 4 to 6 gallons of water to the vessel and add 8 ounces of logwood chips or 4 to 5 ounces of sawdust. Add 2 ounces of fustic chips (or Osage orange or equivalent of other good yellow dye) or 1/2 ounce of extract. Boil this mixture for at least 1 hour.

6. Remove the chips or pour the liquor into another vessel and fill to 4 to 6 gallons with water.

7. Add the wool, heat to the simmer or boil, and maintain this temperature for about 30 minutes.

8. Remove the material and air it well. Add 1/2 ounce of copperas to the dye, stir until dissolved, reenter the wool and dye again for 30 minutes.

9. Remove, drain, squeeze out, air well, wash well with detergent, rinse well, and dry.

Dye Failure Black: All Fibers

In the old days, and at present, blacks have often been produced using a poorly dyed other color as base. For example, I have produced some quite nice reddish blacks from Turkey red failures (slightly unlevel dyeing, etc.). In one such case, logwood and fustic were added, following the recipes for logwood-fustic-madder black. Because the cloth was too red, a deep, solid black did not result, but a nice, level reddish-black emerged. Other examples in which nice blacks may be salvaged from dye failures include the blueing of a poor brown, or vice versa.

Grays: General

Grays may be obtained by using black recipes with greatly reduced quantities of dyestuffs and mordants. If logwood is used, omission of the yellow or yellow and red dye often results in a gray, particularly if a little less logwood is used. The following recipes include my favorite grays.

Steel Gray: Cotton, Linen, or Silk

1. Scour material well, rinse, and leave wet.
2. Mordant with tannin at the rate of 1 ounce per pound or equivalent. Mordant for 6 to 12 hours.
3. Remove, rinse well, and leave wet.
4. Dissolve 1/2 ounce (14 grams or 3 teaspoons) of copperas in 4 to 6 gallons of room-temperature water. When dissolved, work the material for 15 to 30 minutes. With luck, a fine steel gray will result. Wash well, rinse, and dry.

• I often use nitrate of iron (Appendix B) instead of copperas. One ounce (1 1/2 tablespoons) of mordant will probably be about right.

• If the gray is not dark enough try adding a little more copperas or nitrate of iron and rework the material, or work the material in a lime water or washing soda bath for a few minutes (1 teaspoon of washing soda per gallon of water).

• The item may be shaded slightly by soaking the material for a few minutes in weak cochineal liquor. Stronger cochineal will give reddish gray, and stronger yet reddish purple. Addition of madder will give a purple-gray.

Mountain Laurel Gray: Cotton or Linen

1. Produce a gray as in the preceding recipe.
2. Prepare a dyebath by boiling mountain laurel leaves for 2 hours. Use about 2 to 4 gallons of leaves for 1 pound of material, depending upon the degree of gray required. It takes a lot of leaves, but the shades of gray are very nice. Fill the vessel to 4 to 6 gallons with water after the leaves are removed and dye at about 160° F for approximately 1 hour.
3. Remove, wash well with detergent, rinse, and dry.

Prussian Blue–Mountain Laurel Gray: Cotton, Linen, or Silk

1. Dye material a sky-blue shade of Prussian blue (see Blue chapter).
2. Mordant with copperas, about 1/4 ounce (1 1/2 teaspoons) in 4 to 6 gallons room-temperature water, for about 30 minutes.
3. Dye with mountain laurel as in the preceding recipe, about 2 gallons of leaves per pound of material.
4. Finish off as usual.

Sumac Berry Gray: Cotton or Linen

1. Scour material well, rinse, and leave wet.
2. Collect approximately 1 gallon of red staghorn sumac berries and soak overnight in 4 gallons of water.
3. Heat the mix to 180° F and keep at that temperature for 1 hour. Remove the berries.
4. Add the wet, scoured cotton and dye at 180° F for about 45 minutes. Lift material and add about 5 teaspoons of copperas, previously dissolved in hot water.
5. Reenter the goods and dye for an additional 20 minutes.
6. Remove, wash well, and dry.
- If the dyeing is done in an iron pot, less copperas will be required.
- The shade will be affected by the amount of copperas added.
- Some very attractive reddish-purple grays result from this recipe.

Appendix A

Chemicals Used in Traditional Dyeing*

Old or Common Name — Description	Formula
alum (common) — aluminum potassium sulfate	$Alk(SO_4)_2 12H_2O$
ammonia — ammonium hydroxide soln. 5% soln. (develops in old urine naturally from the urea)	$NH_4 OH$
ammonium alum (pickling alum) — aluminum ammonium sulfate	$Al_2(SO_4)_3(NH_4)_2SO_4 \bullet 24H_2O$
aqua fortis — concentrated nitric acid	HNO_3
aqua regia — mixture of concentrated nitric and hydrochloric acids	HNO_3 and HCl
argol (argal or argil) — impure cream of tartar (potassium bitartrate); white or red, depending on whether it is deposited from red or white grapes in wine casks	$KHC_4H_4O_6$
barilla — impure sodium carbonate, originally imported from Spain and the Levant; produced from burned barilla ashes	$Na_2CO_3 \bullet 10H_2O$
black ash — impure potassium carbonate	K_2CO_3
black iron liquor — iron acetate (pyrolignite of iron), produced formerly by dissolving iron filings in vinegar or impure acetic acid	$Fe(C_2H_3O_2)_2$
black lead — impure carbon	C
bleaching powder — chloride of lime, calcium hypochlorite	$CaOCl_2$
block tin — tin cast into ingots or blocks	Sn
blue copperas — copper sulfate	$CuSO_4 \bullet 5H_2O$
bluestone — copper sulfate	$CuSO_4 \bullet 5H_2O$
blue vitriol — copper sulfate	$CuSO_4 \bullet 5H_2O$

*Major references: Adrosko, Napier.

borax – sodium borate $Na_2B_4O_7 \cdot 10H_2O$

brimstone – sulfur S

brown sugar – impure lead acetate, pyrolignite of
lead $Pb(C_2H_3O_2)_2$

calomel (protochloride of mercury) – mercuric
chloride $HgCl_2$

caustic lime lime – calcium oxide CaO

caustic potash – potassium hydroxide KOH

caustic soda – sodium hydroxide $NaOH$

castor oil – castor oil, used in oil mordants for
"Turkey red" chiefly ricinoleic acid

chalk – calcium carbonate $Ca\ CO_3$

chamber-ley – urine ——

chemic – old fashioned bleaching powder, calcium
hypochlorite (Sometimes "chemic" refers to indigo
sulfonate – indigo reacted with sulfuric acid.) $CaOCl_2$

"chrome" – potassium dichromate (first available,
1797) $K_2CR_2O_7$

chrome alum – chromium potassium sulfate
$CR_2(SO_4)_3 \cdot K_2SO_4 \cdot 24H_2O$

cinnabar – mercuric sulfide HgS

common salt (table salt) – sodium chloride $NaCl$

"copper" – copper sulfate $CuSO_4 \cdot 5H_2O$

copperas, blue – copper sulfate $CuSO_4 \cdot 5H_2O$

copperas, green; "iron"; protosulfate of Iron –
hydrated ferrous sulfate $FeSO_4 \cdot 7H_2O$

corrosive sublimate – mercuric chloride $HgCl_2$

cream of tartar – potassium bitartrate $KHC_4H_4O_6$

crystals of tin; "tin"; chloride of tin; (tin mordant for
wool and silk) – stannous chloride $SnCl_2$

double chloride of tin – stannous chloride $SnCl_2$

double muriate of tin – stannous chloride $SnCl_2$

epsom salts – magnesium sulfate $Mg\ SO_4 \cdot 7H_2O$

feathered tin – granulated tin (tin metal is melted in
a ladle and poured into a pail of water from
distance of 4 to 5 feet Sn

flowers of zinc – zinc oxide ZnO

fuller's earth – hydrated magnesium and aluminum
silicates ——

Glauber's salts – hydrated sodium sulfate $Na_2SO_4 \cdot 10H_2O$

glycerine – glycerine or glycerol $C_3H_5(OH)_3$

"good old sig" – old urine, in which ammonia forms NH_4OH

grain tin – metallic tin in prismatic pieces Sn

green vitriol – hydrated ferrous sulfate $FeSO_4 \cdot 7H_2O$

hartshorn – ammonia, ammonium hydroxide NH_4OH

"iron" – ferrous sulfate $FeSO_4$

javelle water – sodium hypochlorite solution $NaOCl$

king's yellow (orpiment), yellow arsenic – arsenic
     trisulfide As_2S_3

lactine – curd of milk – (used in "animalising cotton"
     mordant)   albumen

lemon juice – citric acid (impure) $HOCCOOH(CH_2COOH)_2 \cdot H_2O$

ley – solution, usually of soda ash or pearlash, used
     to scour cotton Na_2CO_3 or K_2CO_3

lime water – aqueous solution, of lime (CaO), calcium
     hydroxide $Ca(OH)_2 \cdot H_2O$

lunar caustic – silver nitrate $AgNO_3$

magnesia nigra – manganese Mn

marine acid – concentrated hydrochloric acid HCl

milk of lime – aqueous solution of calcium hydroxide,
     lime water $Ca(OH)_2 \cdot H_2O$

mineral alkali – soda, soda ash, washing soda Na_2CO_3

"mordant" – generally applies only to aluminum
     acetate made from lead acetate and alum; used in
     cotton and silk mordanting; exists only in solution $Al_2(C_2H_3O_2)_2$

muriates – chlorides; muriate of potassium,
     potassium chloride KCl

muriatic acid – concentrated hydrochloric acid HCl

"nitrate of iron" – ferric sulfate $Fe_2(SO_4)_3$

nitre – potassium nitrate KNO_3

nitromuriate of tin – solution of tin dissolved in
     nitric acid and hydrochloric acids (tin spirits)
           tin nitrates and chlorides

oil of vitriol – concentrated sulfuric acid H_2SO_4

orpiment (yellow) – yellow arsenic, arsenic trisulfide,
     persulfide of arsenic As_2S_3

oxymuriate of potash – potassium chlorate $KClO_3$

oxymuriate of tin – stannic chloride (cotton tin
     mordant) $SnCl_4$

pearlash – purified potassium carbonate (from wood
     ashes; sodium carbonate may be substituted in
     most cases K_2CO_3

permuriate of tin – stannic chloride, perchloride of
    tin (cotton mordant) $SnCl_4$

peroxide – hydrogen peroxide H_2O_2

potash – potassium carbonate K_2CO_3

protochloride of tin, "tin"; tin salts – stannous
    chloride $SnCl_2$

prussic acid – hydrocyanic acid   HCN

purple crystals – potassium permanganate $KMNO_4$

quicksilver – mercury   Hg

realgar – arsenic monosulfide   AsS

red chrome, "chrome" – potassium dichromate $K_2CR_2O_7$

red liquor – solution of aluminum acetate made from
    lead acetate and alum; used in mordanting cotton
    and sometimes silk (The older term "Red Liquor"
    referred to alum mordants used for dyeing red
    with madder.)   ——

red orpiment – arsenic bisulfide As_2S_2

red prussiate of potash – potassium ferricyanide
    (used for production of Prussian blue on wool and
    sometimes silk) $K_6Fe_2(CN)_{12}$

red spirits – solutions of chlorides of tin, used in
    mordanting cotton and wool   ——

Roman vitriol – copper sulfate $CuSO_4 \bullet 5H_2O$

sal ammoniac – ammonium chloride NH_4Cl

saleratus – pearlash (potassium carbonate) over-
    charged with carbonic acid gas K_2CO_3

sali nixon – potassium bisulfate $KHSO_4$

sal prunella – fused potassium nitrate cast into balls
    or cakes KNO_3

sal soda – sodium carbonate, washing soda, soda ash
    (may abe substituted for pearlash) Na_2CO_3

salt cake, Glauber's salts – sodium sulfate $Na_2SO_4 \bullet 10H_2O$

salt of lemons – citric acid $C_2H_2O_3(CH_2COOH)_2 \bullet H_2O$

salt perlate – sodium phosphate Na_2HPO_4

saltpetre – potassium nitrate KNO_3

salt of Saturn – lead acetate $Pb(C_2H_3O_2)_2$

salt of soda – sodium carbonate Na_2CO_3

salt of sorrel – potassium binoxalate $KHC_2O_4 \bullet H_2O$

salt of tartar – potassium carbonate K_2CO_3

salt of vitriol – zinc sulfate $ZnSO_4$

salt sedative – boracic acid, boric acid H_3BO_3

Saxon blue – indigo dissolved (reacted with) concentrated sulfuric acid (an acid dye) $C_{16}H_8N_2O_2(SO_3H)_2$

sig – urine ——

single muriate of tin – half strength of double perchloride of tin, stannous chloride $SnCl_2$

slats of hartshorn – ammonium carbonate $(NH_4)_2CO_3$

soda ash – sodium carbonate Na_2CO_3

soda, calcined – sodium carbonate Na_2CO_3

sour (sour water) – water made acidic by addition of
an acid (definitely sour to taste) ——

"spirits" – solutions of tin chlorides ——

spirits of hartshorn – ammonia, ammonium
hydroxide NH_4OH

spirits of nitre – dilute nitric acid HNO_3

spirits of salt – hydrochloric acid HCl

spirits of wine – ethyl alcohol C_2H_5OH

sugar of lead – lead acetate $Pb(C_2H_3O_2)_2$

sulfate of alumina – aluminum sulfate $Al_2(SO_4)_3 \bullet 18H_2O$

sulfate of indigo – indigo reacted with concentrated
sulfuric acid, indigo sulfonate $C_{16}H_8N_2O_2(SO_3H)_2$

supertartrate of potash – cream of tartar, potassium
bitartrate $KHC_4H_4O_6$

tannic acid – tannic acid, gallotannic acid $C_{76}H_{82}O_{46}$

tannin – impure tannic acid; derived from sumac,
galls, cutch, divi-divi, tara, etc. $C_{76}H_{82}O_{46}$

"tartar" – cream of tartar, potassium bitartrate $KHC_4H_4O_6$

tartar emetic – (tartar emetic used as mordant with
tannin for early basic synthetic dyes) $K(SBO)C_4H_4O_6 \bullet 1/2H_2O$

test blue – sulfate of indigo, indigo sulfonate ——

tincal – borax, sodium borate $Na_2B_4O_7 \bullet 10H_2O$

vegetable alkali – potash, potassium carbonate and
hydroxide K_2CO_3 and KOH

verdigris – copper acetate, basic copper
acetate

$CuO \bullet 2\ Cu\ (C_2H_3O_2)_2$

verditer – copper acetate $CuO \bullet 2\ Cu\ (C_2H_3O_2)_2$

vermillion – mercuric sulfide, "cinnabar" HgS

vinegar – impure acetic acid, usually about 5% CH_3COOH

vitriol – sulfuric acid H_2SO_4

volatile alkali — ammonia, ammonium hydroxide NH_4OH

washing soda — sodium carbonate, sal soda, soda
   ash, soda Na_2CO_3

white copperas — zinc sulfate $ZnSO_4$

white lead — lead carbonate $PbCO_3$

white vitriol — zinc sulfate $ZnSO_4$

white zinc — zinc oxide ZnO

whiting — calcium carbonate, chalk $CaCO_3$

yellow prussiate of potash — potassium ferrocyanide

$$K_4Fe(CN)_6 \cdot 3H_2O$$

Appendix B

Solutions and Mordants Used in Traditional Dyeing

(Caution: Acid-containing solutions and mordants should not be made up and used by persons inexperienced in handling and working with strong acids.)

1. *Acetate of iron (black iron liquor):* Dissolve clean (degreased) iron filings or very small pieces of scrap iron in good, strong cider vinegar. This may take 2 weeks or more. Add an excess of iron, and remove the unreacted iron before using. Keep well stoppered and out of the light. Dilute before using. (Note: A faster method is to dissolve 0000 grade steel wool in a half-and-half mixture of 40 to 80% acetic acid and cider vinegar. Pour the acid carefully into the vinegar— outside, or in a chemical hood. Keep covered. Remove the unreacted steel wool after several days. Dilute by pouring the mordant into water. It is a good iron mordant for cotton, linen, silk, and wool.)

2. *Alum plumb*: Prepare (by soaking and heating) a strong decoction of logwood, and then add to it 1 pound of common alum for each pound of logwood used. (Dye by the one-pot method; no premordanting.)

3. *Barwood spirits* (tin mordant for barwood and sandalwood on cotton and linen): Mix carefully 5 parts of concentrated hydrochloric acid and 1 part of concentrated nitric acid. When cool, add, in small quantities at a time, feathered or powered tin, using 1 ounce of tin for each pound (pint) of mixed acid. Let stand for 1 day or more before using. Keep well stoppered. (Note: Mixing of mordants made from strong acid and metals should be done only in laboratory glassware or heat-resistant glass and outside or in a chemical hood. Always pour acid, carefully, into water, never the reverse. The spirit mordants are concentrated and require dilution before use. Keep all well stoppered, preferably in glass-stoppered bottles or acid reagent bottles, and out of the light. When making dilution, always pour the spirit into water, about 1 gill (4 ounces) per gallon of water. Tin spirits require powdered or feathered tin [No. 5]. Do not inhale tin dust.)

4. *Chloride of iron*: Carefully pour 4 parts by measure of concentrated hydrochloric acid into 2 parts of water. Add degreased iron filings as long as they continue to dissolve. Let settle and pour off the clear liquid. Dilute before use (perhaps 1 ounce of mordant per gallon of water). (Used primarily in the process of dyeing silks and woolens a deep Prussian blue, chloride of iron is better than copperas.)

5. *Feathering tin*: Melt bar tin in an iron pot or ladle and then pour the melt gently from a distance of 3 or 4 feet into a pail of cold water. The metal is reduced to thin flakes and dissolves (reacts) much more quickly when added to acid. (Tin should be melted outside or in a chemical hood, also.)

6. *Fixing solution* (for cotton alum mordants): Mix 1 level-to-heaping tablespoon of disodium hydrogen phosphate, Na_2HPO_4, and 1 level-to-heaping tablespoon of powdered chalk, $CaCO_3$, in 2 gallons of hot (115° to 140° F) water. Work the alum-mordanted cotton or linen in the mixture for a few minutes and let stand for 30 minutes. Rinse well and commence dyeing while still wet.

If only chalk or only phosphate is on hand, make the solution from either, but double the quantity. Sodium arsenate and ammonium carbonate also work. All of these produce a solution of about pH 8. Previous to about 1900 the fixing solution was usually made of sheep or cow dung, 2 to 3 ounces mixed with each gallon of water (Napier, 125). Collect the dung *before* it gets rained on, filter the solution through cheesecloth or old pantyhose before using, and rinse the goods well following treatment. Use just the same as a phosphate-chalk mixture at a temperature of 115° to 140° F.

(Note: Any of the preceding solutions helps fix the alum chemically and, especially, remove unfixed alum. Alum, unfixed—that is, unreacted with the cellulose molecules in the cotton—reacts with the dye, forming a color lake complex which will rub off. If a fixing solution is not used, alum-mordanted cottons should be soaked and rinsed very well several times before dyeing. Alum-mordanted wools should be well rinsed before dyeing as well. About the turn of the twentieth century it was well enough understood that the main effective principles in dung were the calcium and sodium phosphates present. These are in higher concentration in the dung of well-fed animals. John Mercer [mercerization process of cotton] and D. Prince were the first to produce an artificial dung solution from calcium and sodium phosphate solutions in 1839 [Floud, 14].)

7. *Nitrate of iron* (iron mordant for cotton and linen): Pour carefully and slowly 4 parts of concentrated nitric acid into 1 part of

water (outside and away from everything, or in a chemical hood). When cool, slowly add degreased iron filings. Reddish vapor will rise as reaction occurs. Do not inhale this material or permit it to touch vegetation, car paint, etc. Add the iron or steel as long as reaction continues. After letting the solution settle for an hour or so, pour off the dark clear portion and keep in a well-stoppered bottle away from the light. The solution has a dark brownish color and syrupy consistency. It must be diluted by pouring the mordant into water. Amount of dilution depends upon the amount of iron mordanting required. I suggest starting by using 1 ounce of mordant per gallon of water. (Basic ferric sulfate was called "nitrate of iron" and substituted for the preceding [Matthews, 519].)

8. *Plumb spirits* (for logwood purple on cotton and silk): First, mix 6 parts by liquid measure of concentrated hydrochloric acid with 1 part of concentrated nitric acid, and 1 part of water. (Pour the acid *mixture* carefully into the water.) For every pound (pint) weight of the mixture, add by degrees 1 1/2 ounce of feathered or powdered tin. Remember to add the tin in a chemical hood or outside. The spirit should stand a day before being used. (Use this mordant by the one-pot method described next.)

Next, prepare a strong solution of logwood by soaking overnight first, and then simmering for 30 minutes to 1 hour. I suggest about 8 ounces of logwood for each pound of cotton, linen, or silk. Allow the logwood solution to cool and strain out the wood chips. Let sit at room temperature for 1 day. To each gallon of the filtered solution add 1 1/2 pints of the spirit and stir. The preparation will have a pretty plum or violet color, and silk or cotton may be dyed one-pot by this method (no premordanting). The actual dyeing should be done at room temperature for about one hour. The depth of color will depend on the strength of the logwood. (Logwood purples are beautiful, but they will fade badly if subjected to much light.)

9. *Red liquor* (complex aluminum acetate): This is the best alum mordant for cotton, linen, and silk. It is also the most expensive, and, if made with lead acetate, it will produce a poisonous "white lead" by-product which should be disposed of properly. Basic Alum Mordant No. 2 listed in the body of the text is nearly as good. Of the four methods given below, Mairet's method is the latest and probably yields the most mordant out of the materials used.

a. *Red liquor* (Mairet 1939, p. 56): Dissolve 3 pounds of alum (potassium alum) in 1 gallon of warm water. Add, gradually and with stirring, 3 ounces of chalk ($CaCO_3$) made into a thick paste with water. There will be a considerable release of carbon dioxide gas, as from an opened can of beer or pop. Then add, gradually, 2 pounds of lead acetate (sugar of lead) and stir well. Stir occasionally for the next 24 to 36 hours. Let the mixture then rest for 12 hours and carefully decant most of the clear liquid, being careful not to stir up the white lead sediment. Then pour 2 gallons of water on the sediment, stir occasionally for 12 hours, and let rest for 12 more hours. Decant the clear liquid again and add to the first lot. Bottle the mordant tightly or use immediately. When ready to use, add 1 part water to 2 parts of the mordant. (Note: I sometimes add more than 1 part water with good results. This may be necessary because the material must be well covered in order to mordant properly.)

The silk, cotton, or linen should be worked well in the mordant at room temperature for about 10 minutes (after being well wetted out) and allowed to steep for a minimum of 6 hours (12 to 24 hours is better). Wring out, or squeeze in the case of silk, and allow to remain damp for 1 day, and then dry slowly. If extremely brilliant colors are desired, repeat the procedure, using the same mordant. Silk should be dyed as soon after the mordant dries as is convenient, but cotton and linen may be dyed at a later date. Remember to treat the material with the "alum fixing solution" or dung solution before dyeing or following with further mordants. If no fixing solution is available, be sure to wet out (soak for about 6 hours) and rinse the goods very thoroughly before dyeing. Alum acetate mordanted stuffs are hard to wet out; in fact they often appear to be waterproofed. The mordant will last for about 3 weeks, and may be reused during that time. Of course, it becomes weaker with use, and the resulting colors will be less bright.

b. *"Red liquor"* (Napier 1875, p. 350): In 1 gallon of hot water dissolve 2 pounds of potassium alum; dissolve in a separate vessel 2 pounds of acetate of lead in 1 gallon water; in a third vessel dissolve 1/2 pound (I suggest 1/4 pound) soda; mix all the solutions together and stir well for about 10 minutes, then allow to stand overnight; decant the clear solution which is ready for use.

c. *"Red liquor"* (Kuder 1858, p. 198): Dissolve 1 pound of potassium alum in 3 pints of hot water. Add 3/4 pound of sugar of lead, stir for some time, let it set, and afterwards pour off the clear liquid. An extra pint of this mordant may be obtained by pouring a pint of hot water on the sediment, letting it stand, then draining off the clear liquid. If the mordant is not to be used immediately, cork it up in a flask, and put in a cool spot for further use. The sediment is poisonous white lead.

d. *"Red liquor"* (Bancroft, 1794, vol. 1, p. 272): Use calcium acetate instead of lead acetate for production of "acetite of alumina." The white sediment produced by this method is the relatively non-toxic calcium sulfate. After some experimentation, I have settled on the following method):

Dissolve 1 1/2 pounds of potassium alum in 1 gallon of warm water. When cool, add gradually with stirring, 1 1/2 ounce of chalk ($CaCO_3$) made into a thick paste with water. Dissolve 1/2 pound of calcium acetate in 2 quarts of water, and add it, with stirring, to the alum-chalk solution. In about 1 hour decant, carefully, the clear which is the concentrated mordant. Dilute by an equal volume of water if necessary. (Aluminum acetate made from lead acetate produces superior reds. The same mordant made from calcium acetate produces superior clear yellows and oranges.)

10. *"Red spirits"* (cotton and linen tin mordant): Take 3 parts by measure of concentrated hydrochloric acid and 1 part of concentrated nitric acid and mix carefully. Pour this slowly and carefully into 1 part of water. Allow the mixture to cool. Add, in small quantities at a time, 2 ounces of feathered or powdered tin for each pound (pint) of acid. The mordant may be used the next day. Use 1 gill (4 ounce) of this concentrate per gallon of water. This gives a solution of specific gravity of about 1.01 or 2-2-1/2° Tw.

11. *Red spirits* (my modern equivalent): Dissolve 1/2 ounce (14 g) of stannic chloride, $SnCl_4 \cdot 5H_2O$ per quart of water, or 2 ounces per gallon. This gives a solution of specific gravity 1.01 or 2° Tw., which is close to the concentration often called for when tin (red) spirits are required. (White cottons or linens worked in this mordant take on a lemon color as mordanting proceeds. Work and steep the material in the mordant at room temperature for 30 to 60 minutes. The mordant may be reused.)

12. *"Sour"* (*sour water*): To every gallon of water add 1 gill (4 ounces) of concentrated sulfuric acid or 1 1/2 gill of concentrated hydrochloric acid and stir thoroughly. If cotton or linen goods are steeped in a sour, they should be well covered with the liquid. A sour this strong should never be allowed to dry on natural fibers; instead all should be rinsed thoroughly following treatment. (Vinegar diluted with an equal volume of water will give a sour of approximately the same strength as the foregoing.)

Far more dilute sours are often called for. For example, straight indigo-dyed yarns and piece goods should be given, in almost all cases, a sour treatment following dyeing. I partially wash the goods with detergent, rinse, "sour," and then soap or detergent them again and rinse thoroughly. The sour neutralizes any remaining alkali, improves the color, and probably makes the item more fade resistant. A sour for this purpose need only be, to use the older terminology, "slightly sour to the taste." This can be about 1 ounce concentrated sulfuric acid or 1 pint of vinegar to 4 to 6 gallons of water. This should be enough for up to 2 pounds of indigo-dyed cotton, silk, linen, or wool.

Other materials previously used to make "sour water" include diluted lemon juice, fermented bran solution, and even spent cochineal liquor. Two old recipes for bran sour water from Mairet (62) are: "Put 24 bushels of bran in a tub and add about 18 hogsheads of nearly boiling water; acid fermentation soon sets in, and in 25 hours it is ready to use" and "Place several handfuls of bran into hot water and let it stand for 24 hours, or until the water becomes sour. Strain before using."

13. *Test for presence of iron contamination in alum or aluminium sulfate*: Iron contamination in alum will produce a ruinous effect in production of madder and other reds. Test the alum for iron contamination as follows: First, good, iron free alum, when dissolved, should form a colorless solution. Second, you can add a small amount of dissolved potassium ferrocyanide (yellow prussiate of potash). If a blue color develops (Prussian blue), the alum is contaminated with iron and unfit for mordanting cotton or wool to be dyed with reds or yellows. With madder on cotton, iron contaminated alum will give a red-purplish-brown color called "Egyptian purple." (Note: A few drops of potassium ferricyanide solution [red prussiate of potash] may be substituted for the yellow prussiate.)

14. *Cochineal or Lac Scarlet Spirits* (*Napier No. 29*): Carefully pour 9 ounces of concentrated hydrochloric acid into 3 ounces of concentrated nitric acid. Then carefully and slowly pour the mixture into 1 ounce of water. This should be done in strong laboratory glassware outside or in a chemical hood. Do not breathe the fumes.

When cool add, gradually and in small quantities at a time, nearly 2 ounces of feathered or powdered tin. Start by adding a small amount of tin; do not add additional until the first quantity is dissolved (reacted) and so on. This should be done in the hood or in a protected place outside, also. Keep a loose cover on the vessel. The mordant may be used a few hours after the reaction ceases. If there is any sediment, use only the clear. (This recipe is from Napier, 345, and is as good as or better than any other scarlet spirit I have ever used.)

Appendix C

Mordant Chemicals, Approximate Values

Use a scale if available since powdered and granular materials do not weigh exactly the same. One ounce equals 28 1/2 grams.

ONE LEVEL TEASPOON	WEIGHT IN GRAMS	WEIGHT IN OUNCES
alum, K	5.2	2/10
aluminum sulfate	5.0	2/10
chalk (calcium carbonate); or calcium acetate	2.0	1/15
chrome	5.6	2/10
copper sulfate	4.0	3/20
copperas (ferrous sulfate)	3.6	3/20
lead ntirate	10.0	7/20
lye (sodium hydroxide)	5.5	2/10
oxalic acid	3.2	1/10
potassium ferrocyanide or ferricyanide	2.8	1/10
soda ash (sodium carbonate)	5.0	2/10
tannic acid	3.0	1/10
tannin (fluffy)	2.0	1/14
tartar (cream of tartar)	3.0	1/10
tin (stannous or stannic chloride)	5.8	2/10

Glossary

acid A chemical compound, with a sour taste, that releases hydrogen ions when dissolved in water. The strength of the acid depends upon the concentration of hydrogen ions. Acids react with bases, forming salts. Acids have a pH less than 7. Acids can be organic — such as citric, tartaric, tannic, oleic, and acetic — and inorganic or mineral — such as sulfuric, hydrochloric, nitric, and phosphoric.

adjective dye A natural dye that requires one or more mordants for effective dyeing.

alkali A chemical compound, with a bitter taste, that releases hydroxyl ions in solution. The strength of the alkali depends upon the concentration of hydroxyl ions. Alkalis react with acids, forming salts, and have a pH greater than 7. Common alkalis include ammonia, sodium and potassium hydroxides, and carbonates.

base Same as alkali.

bleach To whiten by exposure to sunlight (ancient) or by chemicals such as hydrogen peroxide and sodium hypochlorite (modern).

bistre Another name for manganese bronze.

bottom color In a compound color, the first color dyed. For example, a good black results from dyeing black walnut brown on a deep indigo blue bottom.

buffers Chemical compounds that prevent appreciable pH change of a solution when either an acid or a base is added.

cellulosics Fibers composed of cellulose. Cellulose is composed of polymers of simple sugar (glucose) molecules. Cellulosic fibers include all of the vegetable fibers, such as cotton, linen, ramie jute, hemp, and fibrous basketry materials.

chemical dyes (synthetic dyes) All dyes, natural or synthetic, are chemicals, but the term "chemical dyes" refers to those synthesized in the laboratory.

color (hue) Red, blue, orange, etc.

   Intensity or saturation Depth or brilliance of a color.

value The degree of lightness or darkness of a color when compared with black or white.

shade A color diluted with darker than middle-value gray or black. In older terminology, shade also refers to intensity of a color, e.g., dark or light blue.

tint A color, such as red, diluted with white.

tone A color diluted with less than middle-value gray.

crock Dyes that rub off (are not *rubfast*).

dyestuff or dye Natural or traditional dyes are organic chemicals produced by plants or animals, such as indigo, and inorganic chemical compounds, such as iron oxides.

exotic dyes Imported natural dyestuffs such as madder, cochineal, logwood, and natural indigo.

false dyes Dyes such as turmeric that fade rapidly upon exposure to sunlight.

fancy colors Bright flashy colors, usually made from natural dyes liable to fade, such as "fancy brazilwood reds."

fast colors (or dyes) Dyes that do not fade appreciably when exposed to light (*lightfast*); dyes that are resistant to washing (*washfast*); dyes that are fast to rubbing (*rubfast*); and dyes that are resistant to sweat (*sweatfast*). Some dyes are lightfast but not washfast, etc.

felt Development of a nonwoven fabric by wetting, heating, and agitating woolen fibers.

fix To effect more permanent chemical combination.

fixing solution A solution composed of a weakly alkaline substance, such as calcium carbonate, that facilitates more permanent chemical combination of an alum mordant to cellulosic fibers.

flower Nonreduced indigo that collects on the surface of zinc-lime and ferrous sulfate indigo vats.

fugitive (dyes) Dyes that rapidly fade when exposed to light.

killing Dissolving (reacting) a substance in an acid, as iron filings in hydrochloric acid; killing iron.

lake (color lake) The insoluble mordant-dye complex that is chemically combined with the fiber in a mordant dyed item. Color lakes are produced with adjective dyes. Color lakes produced by reacting a dye with a metallic salt, such as madder (alizarin) with alum, were also used in inks and paints.

leveling agent A material, such as Glauber's salts, added to a dyebath to facilitate even or level dyeing.

ley Solution of an alkali, used for scouring.

lint   Ginned, but unspun cotton fiber.

litmus (litmus paper)   A dyestuff extracted from archil lichens that turns red in acidic solutions and blue in alkaline solutions.

modifier   A chemical added to the dyebath to alter the color, e.g., alkali added to a logwood dyebath produces a darker blue; tin added to a yellow dyebath produces a brighter yellow.

mordants   Chemicals (metal oxides, tannins, and oxyfatty acids) that are necessary to chemically fix (make washfast) most natural dyestuffs. The mordant combines both with the dye molecule and the fiber molecule, producing a permanently fixed insoluble color lake.

mordant assistant   A chemical that increases the effectiveness of a mordant. Examples are cream of tartar and washing soda.

nonreactive   A substance, usually a dye vessel, that does not react chemically and is not affected by dyebaths, mordants, acids, or alkalis.

overdye   To dye one color over another to produce a compound color. For example, indigo blue overdyed with a yellow produces green.

pH   A measure of the acidity, alkalinity, or neutrality of a solution. pH O–6.5 is acidic, 7 is neutral, and 7.5–14 is alkaline. The system is logarithmic, so that a solution of pH 1 is 10 times as acidic as pH 2 and 100 times as acidic as pH 3.

protein fibers   Wool and silk are the protein fibers. Proteins are polymers of amino acids.

saddening agent   Addition to the dyebath of iron mordant, which ultimately dulls the color. Dyeing in an iron vessel has the same effect, particularly on bright colors.

scald   (scalding temperature) 150° F or 65° C.

sharpen   Addition of lime or zinc dust, or both, to an indigo vat that is reverting to an unreduced state.

simmer   A temperature of 180° to 190° F or 82° to 88° C.

slurry   A thin, watery mixture.

sour   Treatment with an acidic solution, usually weak sulfuric or acetic acid.

stain   The effect produced by an adjective dye on unmordanted fiber or by a colored non dyestuff. Stains are rarely washfast.

substantive Dye   A natural dye that dyes (affects chemical combination with natural fibers) without added mordants. The dyestuff may contain its own natural mordant (tannins) as do black walnut hulls, or it may not, as is the case with turmeric, annotto and saffron.

tender To become weakened. Cellulosic fibers are tendered by mineral acids.

true dyes Natural dyes that are quite lightfast, such as madder reds and indigo blues.

vat dyes Indigo and Tyrian purple. These dyes must be reduced chemically to a leuco form in an alkaline dyevat prior to dyeing.

work Manipulation of yarn or piece goods in a mordant or dyebath to facilitate even and thorough penetration.

References

A Practical Treatise on the Arts of Dyeing and Calico Printing. (By an Experienced Dyer). 1846. New York: Harper.

Adrosko, R. 1971. *Natural Dyes and Home Dyeing*. New York: Dover.

Baker, J. R. 1958. *Principles of Biological Microtechnique*. New York: Wiley.

Bancroft, E. 1814. *Experimental Researches Concerning the Philosophy of Permanent Colours*. 2 vols. Philadelphia.

Bemis, E. [1806] 1973. *The Dyer's Companion*. New York: Dover. (Unabridged reprint of the 2d. [1815] edition, published by E. Duyckinck, New York.)

Blue Traditions. 1973. (Introduction and commentary by D. S. Katzenberg). Baltimore: The Baltimore Museum of Art.

Bliss, A. 1981. *A Handbook of Dyes from Natural Materials*. New York: Scribner's.

Blumrich, S. 1983. "Indigo-The Devil's Dye." *Surface Design Journal* 7 (3):18–23.

———. 1984. "Indigo Resist Printing in Scheessel, Germany." *Surface Design Journal* 8(4):19–24.

Bronson, J. and R. [1817] 1977. *Early American Weaving and Dyeing*. New York: Dover. (Unabridged and slightly corrected reprint of the original edition published by William Wilkins, Utica, N.Y.)

Brunello, F. [1968] 1973. *The Art of Dyeing in the History of Mankind*. Trans. Bernard Hickey. Vicenza: Neri Pozza Editore. (English translation made available by the Phoenix Dye Works, Cleveland, Ohio.)

Bryan, N. G. [1940] 1978. *Navajo Native Dyes, Their Preparation and Use*. Palmer Lake, Colo.: The Filter Press. (Originally Published by the U. S. Bureau of Indian Affairs.)

Buchanan, R. 1987a. *A Weaver's Garden*. Loveland, Colo.: Interweave Press.

———. 1987b. "Grow Your Own Colors – Plant A Dye Garden." *Spin-Off* 11(1):35–40.

Cain, J. C. and J. F. Thorpe. 1933. *The Synthetic Dyestuffs and the Intermediate Products from which They are Derived*. London: Charles Griffen.

Caley, E. 1927. "The Stockholm Papyrus." *Journal of Chemical Education* 4(8):979–1002.

Casselman, K. L. 1980. *Craft of the Dyer: Colour from Plants and Lichens of the Northeast*. Toronto: Univ. of Toronto Press.

Ciba Review. "Medieval dyeing," no. 1, (1937); "India, its Dyers," no. 2 (1937); "Purple," no. 4 (1937); "Scarlet," no. 7 (1938); "Dyeing and Tanning in Classical Antiquity," no. 9 (1938); "Trade Routes and Dye Markets in the Middle Ages," no. 10 (1938); "Weaving and Dyeing in Ancient Egypt nd Babylon," no. 12 (1938); "Great Masters of Dyeing in eighteenth Century France," no. 18 (1939); "Madder and Turkey Red," no. 39 (1941); "Indigo," no. 85 (1951); "Identification of Colourants in Ancient Textiles," no. 8 (1963); "Textiles in Biblical Times," no. 2 (1968).

Colton, Mary-Russell Ferrell. 1965. *Hopi Dyes*. Flagstaff: The Museum of Northern Arizona Press.

Conley, E. 1957. *Vegetable Dyeing*. Penland, N.C.: Penland School of Handicraft.

Cooper, J. 1815. *A Practical Treatise on Dyeing and Calico Printing*. Philadelphia: William Fry.

Crews, P. 1981. "Considerations in the Selection and Application of Natural Dyes: Mordant Selection." *Shuttle, Spindle and Dyepot* 12 (2):15 and 62.

Davenport, E. G. 1955. *Your Yarn Dyeing*. Rockville, Md: Craft and Hobby Book Service.

Davidson, M. F. 1974. 2d ed. *The Dye Pot*. Gatlinburg, Tenn.: privately printed.

Davis, Virginia. 1987. "William Morris: Discharge and the Art of Dyeing." parts 1 and 2. *Surface Design Journal* 12(2):24–26 and 12(3): 23–24.

Dick, W. B. 1974. *Dick's Encyclopedia of Practical Receipts and Processes. Or How They Did it in the 1870's*. New York: Funk and Wagnalls.

Dye Plants and Dyeing – A Handbook. 1964. Brooklyn, N.Y.: Brooklyn Botanic Garden.

Edelstein, S. M. 1954. "The Dual Life of Edward Bancroft." *American Dyestuff Reporter*" 43(22):712–35.

———. 1972. *Historical Notes on the Wet-Processing Industry*. New York: Dexter Chemical Corporation.

Ellis, Asa. 1798. *The Country Dyer's Assistant.* Brookfield, Mass.

El Neuvo Constante. 1981. Baton Rouge: Div. of Adm., Adm. Serv.

Floud, P. 1961. "The English Contribution to the Chemistry of Calico Printing Before Perkin." *CIBA,* no. 18(1961).

Furry, M. and B. Viemont. 1935. *Home Dyeing with Natural Dyes.* Bureau of Home Economics, No. 230. Washington, D.C.: U.S. Dept. of Agriculture.

Gerber, F. 1977. *Indigo and the Antiquity of Dyeing.* Oak Ridge, Tenn.: privately printed.

———. 1978a. *Cochineal and the Insect Dyes.* Oak Ridge, Tenn.: privately printed.

———. 1978b. *The Investigative Method of Natural Dyeing, and Other Dye Articles.* Oak Ridge, Tenn.: privately printed.

———. 1983. "The Chemistry and Use of Indigo: The Mystery Revealed." *Surface Design Journal* 8(2):23–27.

———. 1984a. "Indigo Recipes." *Surface Design Journal* 8(3):14–17.

———. 1984b. "Color: Usage." Parts 1 and 2. *Surface Design Journal* 9(3):7–9 and 10(1):7–10.

Gerber, F. and Willie Gerber. 1974. "Quercitron, the Forgotten Dyestuff, Producer of Clear, Bright Colors." *Shuttle, Spindle, and Dyepot* 5(1 and 2): 25 and 87.

Gittinger, M. 1982. *Master Dyers to the World.* Washington, D.C.: The Textile Museum.

Goodrich, F. L. 1931. *Mountain Homespun.* New Haven: Yale University Press.

Grae, J. 1974. *Nature's Colors.* New York: Macmillan.

Grierson, S. 1986. *The Colour Cauldron.* Perth, Scotland: privately printed.

Gwynne, E. 1982. "Chemical Dyeing of Linen." *The Weaver's Journal* 7(2):23–26.

Hellot, M., M. Macquer, and M. LePileur D'Apligny. 1789. *The Art of Dyeing Wool, Silk, and Cotton.* (Translated from the French). London. R. Baldwin.

Hochberg, B. 1983. "Events in Textile History." *Fiberarts* 10(2): 52.

Holding, M. 1949. *Notes on Spinning and Dyeing Wool.* 4th ed. London: Skilbeck Brothers.

Horsfall, R., and L. Laurie. 1946. *The Dyeing of Textile Fibres.* London: Chapman and Hall.

Hummell, J. J. 1885. *The Dyeing of Textile Fabrics.* London: Cassel.

Indigo in America. 1976. Parsippany, N.J.: BASF Wyandotte Corp.

Joslyn, C. 1986. "Cloth and Surface Design in Ivory Coast, West Africa." Parts 1 and 2. *Surface Design Journal* 10(2):11–15 and 10(3):23–5.

Journal of the Chicago Horticultural Society. 1976. *Plant Dyes and Natural Dyeing*, vol. 3, no. 1.

Kajitani, N. 1979. *Traditional Dyes in Indonesia. Irene Emery Roundtable on Museum Textiles - Indonesian Textiles*. Washington, D.C.: The Textile Museum.

Kierstead, S. P. 1950. *Natural Dyes*. Boston: Brandon Press.

Kolander, C. 1978. "Natural Dyeing of Silk Fiber." *Shuttle, Spindle and Dyepot* 9(2):58–60.

Kuder, S. 1858. *The Practical Family Dyer*. Allentown, Penn.: Blumer Leisenring.

Leggett, W. 1944. *Ancient and Medieval Dyes*. New York: Chemical Publishing.

Liles, J. N. 1982. "The Mineral Dyes." Parts 1 and 2. *Shuttle, Spindle and Dyepot* 51:54-57 and 52:60-63.

———. 1984. "The Mineral Dyes." In *Dyeing for Fibres and Fabrics*, 62-64. St. Lucia: The Australian Forum for Textile Arts, University of Queensland. (2nd ed., 1987, pp. 85-88.)

———. 1985. "Dyes in American Quilts Made Prior to 1930, with Special Emphasis on Cotton and Linen." In *Uncoverings 1984*, 29–40, Vol. 5 of the Research Papers of the American Quilt Study Group, Mill Valley, California.

———. 1988–89. "On Taking Things for Granted." *Patchwork Patter* 16(3):10; 16(4):6–7; 17(1):16–17.

Liles, J. N. and M. P. Liles. 1984a. "Bancroft's Mordant." *Shuttle, Spindle and Dyepot* 15(3):76–79.

———. 1984b. "Bancroft's Mordant: A Useful One-Pot Natural Dye Technique." In *Dyeing for Fibres and Fabrics,* 70–72. St. Lucia: The Australian Forum for Textile Arts, University of Queensland.

Mairet, E. M. 1939. *Vegetable Dyes*. New York: Chemical Publishing.

Manual for the Dyeing of Cotton and Other Vegetable Fibres. 1936. New York: General Dyestuff Corporation.

Matthews, J. M. 1920. *Application of Dyestuffs to Textile, Paper, Leather and Other Materials*. New York: Wiley.

Mayer, F. and A. Cook. 1943. *The Chemistry of Natural Coloring Matters*. New York: Reinhold Publishing Corporation.

Miller, D. 1984. *Indigo from Seed to Dye*. Aptos, Calif.: Indigo Press.

Molony, C. 1834. *The Modern Wool Dyer*. Lowell, Mass.

Napier, J. N. 1875. *A Manual of Dyeing and Dyeing Receipts*. London: Charles Griffin and Co.

Natural Plant Dyeing. 1973. Brooklyn, N.Y.: Brooklyn Botanic Garden.

Partridge, W. 1847. *A Practical Treatise on Dyeing Woolen, Cotton, and Silk*. New York: Wm. Partridge's Son and Co.

Pellew, C. E. 1912. "The Dyestuffs of the Ancients," *Handicraft* 5(5): 63–71 and "The Dyestuffs of our Ancestors," *Handicraft* 5(6):87–96.

———. 1928. *Dyes and Dyeing*. New York: Robert M. McBride.

Pettit, F. 1974. *America's Indigo Blues*. New York: Hastings House.

Polakoff, C. 1980. *Into Indigo*. Garden City, N.Y.: Anchor Press/Doubleday.

Ramsey, B. 1983. "Cotton Country: Redbud Georgia, 1873-1907." In *Quilt Close-Up – Five Southern Views*. Chattanooga: The Hunter Museum of Art.

Robinson, S. 1969. *A History of Dyed Textiles*. Cambridge: M.I.T. Press.

Rosetti, G. [1548] 1969. *The Plichto of Instructions in the Art of the Dyers*. (Translation of the first edition of 1548 by Edelstein and Borghetty). Cambridge: M.I.T. Press.

Rossi, G. 1986. "Laran Artists From the Mountain's of China's Guizhou Province." Parts 1 and 2. *Surface Design Journal* 10(3):18–22 and 10(4):20–22.

———. 1988. "Growing Indigo in China's Guizhou's Wanchae District." *Surface Design Journal* 12(4):28–29.

Sandberg, G. 1989. *Indigo Textiles, Technique and History*. London: A&C Black; Asheville, N.C., Lark Books.

Schaefer, C. 1941. "History of Turkey Red Dyeing." *Textile Colourist* 63:465–90.

Schweppe, H. 1976. "Testing of Old Textile Dyeings." *The B.A.S.F. Review*, no. 26, pp. 29–36.

Seymour, J. 1984. *The Forgotten Arts*. London: Dorling Kindsedy.

Soderberg, B. 1973. *Color from Plants*. Hollywood, Calif.: privately printed.

Smith, D. 1847. *The Dyer's Instructor*. London: Simkin and Marshall.

———. [1897] 1900. *The Dyer's Instructor*. 4th ed. Philadelphia: Henry Carey Baird and Co.

Stanfield, N. et al. 1971. *Adire Cloth in Nigeria*. Nigeria: The Inst. of African Studies, Univ. of Ibadan.

Thurstan, V. 1977. *The Use of Vegetable Dyes*. Leicester, Eng.: Dryad Press.

Tidball, H. 1965. *Color and Dyeing*. Shuttle Craft Monograph 16. Freeland, Wash: HTH Publishers.

Trotman, E. 1970. *Dyeing and Chemical Technology of Textile Fibres*. London: Charles Griffin and Co.

Tucker, W. 1822. *The Family Dyer and Scourer*. London.

Viner, W. S. and H. E. S. Viner. 1946. *The Katherine Pettit Book of Vegetable Dyes*. Saluda, N.C.: Excelsior Press.

Weigle, P. 1974. *Ancient Dyes for Modern Weavers*. New York: Watson-Guptill.

Wipplinger, M. 1985. "Dyes of Mexico." *Spin-Off* 9(3):32–35.

———. 1985. "Art of Silk Dyeing." *Weaver's Journal* 10(2):46–52.

Wright, N.P. 1963. "A Thousand Years of Cochineal: A Lost but Traditional Mexican Industry on Its Way Back." *American Dyestuff Reporter* 52, no. 17.

Index

34, 167, 169; in Prussian blue,
49
"iron liquors," 24, 201
iron mordants, 24, 26, 29–30
iron-tannates, 184
"iron tendering:" in blacks and
browns, 187
Isatis tinctoria, 54, 99

Juglans cinera, 174
Juglans nigra, 174
juglone, 174

kermes, 102, 156, 167
khaki, 181

lac, 102, 138, 156, 167
lead chromate, 34, 146
leucoindigo, 54
lime water, 94–95
liver brown, 182
logwood: blacks, 184; blues, 43,
45–47; greens, 148–49; history,
43–45; purples, 25, 156, 158–59
Lonchocarpus cyanescens, 54, 99
London smoke, 182

Maclura pomifera, 35, 39
madder: cultivation, 104; history,
103–5; species used, 103
madder browns, 174, 178–79, 182
madder oranges, 167–69
madder pink, 123
madder purples, 156–58, 162–63
madder reds: procedures, 105–6;
cotton, 106–11; silk, 123–24;
wool, 126
manganese bronze, 174, 179–80
marigolds, 34
Marsdenia sp., 54
mauve, 2
milkweed, 54
mimosa, 34, 39
mineral dyes, 4
mordants: general features and
chemistry, 4; cotton and linen,
18–26; silk, 26–27; wool, 27–31;
traditional, Appendix B

Morinda augustifolia, citrifolia,
and *tinctoria*, 103
Morus tinctoria, 34
Murex sp., 155

Napoleon's blue, 47, 51
natural dyes: how good?, 3
Nerium tinctorium, 54

"old fustic," 34, 37, 146
Oldenlandia umbellata, 103
oleander, 54
onion skins, 35
orange dyes: history, 167
orpiment: in direct indigo dyeing,
92
Osage orange, 35, 37, 39

painting, 4
Paris blue, 47
Paris mud, 182
peach leaves, 35, 39
pencil blue, 93–94
Perkin, William H., 2
"Persian berries," 33, 146
pokeberries, 103
Polygonum persicaria, 35, 39
Polygonum tinctorium, 54, 99
protein fibers, 4
Prunus persica, 35, 39
Prussian blue: discharge, 50; dyes,
49–53; history, 47; in green,
146, 150, 151; in tie dye work,
50; test for iron, 9
pseudo-purpurin, 104
puce, 159–60
purple dyes: history, 155
Purpura sp., 155
purpurin, 104

Queen Anne's lace, 35, 38
quercitron, 33, 35, 37
Quercus velutina, 33, 35

Raymond's blue, 47
Red dyes: history, 102
"red liquor," 22, 203–5
"red spirits," 25, 205